Current Trends in the Theory of Fields

(Tallahassee-1978)

A Symposium in Honor of P.A.M. Dirac

"It has happened sometimes that ideas have come to me while taking a walk."

AIP Conference Proceedings
Series Editor: Hugh C. Wolfe
Number 48
Particles and Fields Subseries No. 15

Current Trends in the Theory of Fields

(Tallahassee-1978)

A Symposium in Honor of P.A.M. Dirac

Editors
J.E. Lannutti and P.K. Williams
Florida State University

American Institute of Physics
New York 1978

L.C. Catalog Card No. 78-72948
ISBN 0-88318-147-9
DOE CONF-780495

After the banquet.

"My famous brother-in-law"

Emin T
T = L
CTL

PREFACE

The Jubilee Conference "Current Trends in the Theory of Fields" in honor of Nobel Laureate Paul A.M. Dirac was held at Florida State University, Tallahassee, Florida, April 6-7, 1978.

The occasion marked the 50th anniversary of the publication of the Dirac Equation and was attended by about 75 physicists from around the world.

The conference was organized into three sessions, each with two or three principal speakers and a panel discussion with additional short talks by other speakers, which are included herein. Unfortunately, we are not able to include the spontaneous parts of the discussions, nor the delightful after-dinner speech by A. Pais given at the banquet.

The organizing committee was chaired by J. E. Lannutti and included S. Glashow, S. Drell and J. Wheeler in addition to local people (P.K. Williams, V. Hagopian, L. Halpern, J. D. Kimel and D. Robson). The conference drew support from the President's Club-Florida State University, The U. S. Department of Energy and the National Science Foundation.

We gratefully acknowledge the help of the FSU Office of Continuing Education. Special thanks are due to the members of the Physics Department who helped, notably Dr. J. F. Owens, A. Thibault and A. Connor.

July, 1978

J. E. Lannutti
P. K. Williams

TABLE OF CONTENTS

NON-ABELIAN GAUGE THEORIES AND QUARK CONFINEMENT

G. 't Hooft
Institute for Theoretical Physics
University of Utrecht, The Netherlands

ABSTRACT

After some philosophical remarks concerning the development of theoretical physics in this century a short introduction to non-Abelian gauge theory is given. If we concentrate on the case where the local gauge group is SU(N) we find that gauge-invariant definitions can be given of the concepts of electric and magnetic flux. Both are quantized and conserved modulo N. Either the electric or the magnetic flux must be squeezed in narrow flux tubes called vortices or strings. If the vortices are electric then quarks are confined into grouplets of multiples of N quarks. A phase transition separates the "quark confinement mode" from the "Higgs mode", the latter being easier accessible by the use of perturbation expansion.

INTRODUCTION

It is a great honor for me to speak here at the celebration of the anniversaries of a great physicist and a beautiful equation. I choose as my subject a tantalizing problem in present-day quantum field theory and will discuss some ideas that bring us close to its resolution. How do these ideas connect with the fundamental developments in physics early this century?

Let me summarize what, in my opinion, were the most basic discoveries made in the physical sciences. In previous centuries it was already discovered that progress could be made by performing experiments, and using objectivity rather than prejudice to determine what the laws of physics are. Thus the laws of classical mechanics and chemistry were settled and atomic physics was well on its way. The fathers of this idea were Gallileo, Newton, Huijgens, etc.

The most striking event in the beginning of this century was the development of both quantum mechanics and the theories of special and general relativity. Perhaps their near coincidence was not an accident. Although these theories look very different they are all linked to one basic new discovery: the dogma concerning the role of the observer in an experiment. In interpreting the experiment it may be crucial to realize that the observer is part of the physical system he is dealing with. Thus physical laws govern the functioning of the clocks he uses, and the photons he emits to look at a physical system may disturb the same system. It may not be possible to use a preconceived universal reference frame. So we put Einstein, Planck, Heisenberg, ... under just one heading.

Finally we have the discovery of mathematics and logics, which is being done over and over again at higher levels of sophistication and abstraction. To discover a physical law one must start with a large set of possibilities and then try to reconcile several physical requirements which might seem contradicting at first sight. The

ISSN: 0094-243X/78/001/$1.50

Dirac equation is an excellent example of an achievement made along these lines.

No truly great discovery has been made since. We still make use of only these three philosophies when dealing with the problems of our times. For example we have been facing the problem how to construct a working field theory for vector-like forces of various kinds as they seemed to govern weak, electromagnetic and even strong interactions. The answer turned out to be "non-Abelian gauge theory", originated by Yang, Mills[1] and Shaw[2] who were obviously inspired by Einstein's general relativity theory. It was much later however that the necessity of this gauge principle was shown to follow from physical requirements such as renormalizability and unitarity.

I now come to my real subject: the elusive quarks. We are trying to make a theory for particles that can never be isolated completely but nevertheless seem to be fundamental building blocks of matter. Naively, we can here apply what we learned in lesson # 2, concerning the role of the observer. We should take his readings litterally ("I do not observe any free quarks") and refrain from the use of a subjective, universal reference frame (the concept of particles making up for all matter). Conclusion: quarks are silly, and attempts to make a quark theory are as futile as those for the ether theory etc.

I argue that the above reasoning is shortsighted and wrong for various reasons. First of all "observation" is confused with "isolation as a free particle". Quarks have been observed numbers of times but always in the direct vicinity of other quarks. Secondly, the idea of permanently confined but otherwise normal particles[3] should be abandoned only if it leads to real contradictions or if a better alternative theory exists. For instance one could try to introduce quarks as objects with wrong statistics, or magnetic monopoles, or simply avoid any notion of quarks altogether. And then we should apply lesson # 3: do the mathematics and consider the physical requirements. If "asymptotic freedom"[4] is taken as one of the requirements one naturely selects out the truly confined "colored" but otherwise ordinary fermionic quarks as the only acceptable theory.

Selected by elimination. But does the theory work? My claim is yes. Confinement, by infinitely rising linear potentials, can be a property of the vacuum state precisely as basic as "superconductivity" of a solid at low temperatures. It is a new phase, a new appearance, for a system of fields, separated from another phase ("Higgs mode") by a phase transition at a critical value of certain parameters. Technical details will have been published[5] by the time these notes appear in print.

NON-ABELIAN GAUGE THEORY

The field equations in quantum-electrodynamics are invariant under an Abelian group of gauge transformations:

$$\begin{aligned} \psi &\to e^{i\Lambda(x)}\psi \quad , \\ A_\mu &\to A_\mu - \frac{1}{e}\,\partial_\mu \Lambda \ . \end{aligned} \qquad (1)$$

The gauge group is U(1) and since Λ is space-time dependent we call this a local invariance.

In the non-Abelian extension of this theory the fermion field is replaced by a multiplet of fermion (or scalar) fields and one demands invariance under a larger (non-Abelian) group of transformations[1]:

$$\psi \rightarrow \Omega(x)\psi \ . \tag{2}$$

The vector potential field then must be replaced by a multiplet of vector fields forming a Hermitean matrix. They must transform as

$$A_\mu \rightarrow \Omega(x)\ A_\mu \Omega^{-1}(x) + \frac{1}{gi}\ \Omega(x)\ \partial_\mu \Omega^{-1}(x) \ . \tag{3}$$

The generalization of the electromagnetic fields $F_{\mu\nu}$ is

$$G_{\mu\nu} = \partial_\mu A_\nu - \partial_\nu A_\mu + ig[A_\mu, A_\nu] \ . \tag{4}$$

The inhomogeneous Maxwell equations are

$$D_\mu G_{\mu\nu} \equiv \partial_\mu G_{\mu\nu} + ig[A_\mu, G_{\mu\nu}] = j_\nu \tag{5}$$

where j_ν is the current of the fermion and/or scalar fields.

NON-ABELIAN ELECTRIC FLUX

The extra commutators in eqs. (4) and (5) are special features of the non-Abelian theory. They give more non-linear terms in the field equations. Replacing the commutator in (5) to the right hand side we notice that there is an extra current due to the gauge field themselves, i.e., the gauge quanta are charged, contrary to the photons in the Abelian case. This implies that, even without fermions or scalar particles the gauge-electric field is not conserved in the usual sense. Is there any way in which we can restore the concept of a conserved electric flux in a pure gauge theory without additional fermions or scalars?

The answer to that question depends on the gauge group chosen. We will consider SU(N) where N is as yet a free integer $\geqslant 2$. (For the strong interaction theory with quarks we will have $N = 3$.) These groups have an invariant Abelian subgroup Z(N), whose elements are, if written as N×N matrices,

$$\Omega_n = e^{2\pi i n/N}\ I \tag{6}$$

where I is the identity matrix, and n is an integer, $0 \leqslant n < N$. Obviously, the gauge quanta generated by the field A_μ are invariant under Z(N) transformations (see eq. (3)). They therefore have no Z(N) "charge". We could introduce spectator fermions ψ, for instance as elementary N-fold multiplets, which do transform under Z(N). They would carry a special gauge field configuration around them that can never be neutralized by gauge quanta. This allows one to define the concept of a conserved electric flux.

However this flux differs from ordinary electric flux in two important ways. Firstly, the flux is quantized, because the smallest unit is the amount emitted by just one N-multiplet. Secondly, if we put N of these smallest units together then we obtain a source which is again singlet under Z(N), so that its effect can be neutralized by the gauge quanta. Consequently, the conservation is only modulo N. If Φ is the integer that counts this electric flux then a more convenient symbol is

$$B = e^{2\pi i\Phi/N} \tag{7}$$

which is then multiplicatively conserved.

Now we will define[5] an operator B(C) for each closed curve C that measures (multiplicatively) the amount of Z(N) electric flux that goes through the curve C. Its eigenvalues are $e^{2\pi in/N}$. At any point x not on C our operator B(C) will be just a gauge transformation given by some $\Omega(x)$. Since we only need to tell how B(C) acts on the gauge field we need only specify Ω up to elements of Z(N). We now choose Ω such that if x is followed along a contour C' that winds through C exactly once, labeled by a parameter $o \leqslant \theta < 2\pi$, then

$$\Omega(\theta=2\pi) = e^{2\pi i/N}\,\Omega(\theta=o) \ . \tag{8}$$

Not only does this operator have the desired multiplicative properties and $[B(C)]^N = I$ (since physical states are invariant under single-valued gauge transformations*), but one can also verify that in the presence of an elementary "spectator" N-fold multiplet particle the flux changes by one unit if the particle goes through the curve C.

NON-ABELIAN MAGNETIC FLUX

We have a similar difficulty when we try to generalize the concept of magnetic flux to the non-Abelian case. The Abelian formula was simple: the flux through a closed curve C is

$$\Phi = \oint_C A_i dx^i \ . \tag{9}$$

But in the non-Abelian case this is not gauge-invariant, therefore not an observable quantity. A gauge invariant operator, defined in terms of the vector potential field on a closed curve, does exist:

$$A(C) = \mathrm{Tr}\ P \exp[ig \oint_C A_i dx^i] \ . \tag{10}$$

Here P stands for path ordering of the exponent the integral. If we write

*The necessaty of an infinitesimal smearing of the infinities arising on the curve C might cause complications here which we will ignore.

$$A(C) = N\, e^{2\pi i\Phi/N}$$

then Φ is only defined modulo N, just as in the electric case. The eigenvalues of A(C) are not so restricted as in the electric case, except when the curve C is in a vacuum: on C we have then

$$A_\mu = \frac{1}{gi}\,\Omega(x)\;\partial_\mu\Omega^{-1}(x).$$

If we do not insist on having a vacuum at the interior of C then this Ω may be multivalued, so that we obtain

$$A(C) = N\, e^{2\pi in/N}\ , \tag{11}$$

and then Φ is integer modulo N.

Thus notice that the dual similarity between electric and magnetic fields is still there in the non-Abelian theory. They both are conserved modulo N, and must be integers. We further note that the operators A(C) and B(C') do not commute (just as in the Abelian case). One finds:

$$A(C)\; B(C') = B(C')\; A(C)\; e^{2\pi in/N}\ , \tag{12}$$

where n is now the number of times that the curve C' winds through C.

THE HIGGS MODEL: NIELSEN-OLESEN VORTICES

Properties of the operators A and B can be studied in a soluble model. With soluble I mean that perturbation expansion can be applied and converges sufficiently rapidly so that the results can be trusted. Pure gauge theories are not (yet?) soluble in this sense because of infrared divergences. But we can add scalar field multiplets H_i that transform under gauge transformations the way the Hermitean matrices A_μ do:

$$H_i' = \Omega\; H_i\; \Omega^{-1}\ . \tag{13}$$

And choose the self-interactions of H such that for instance

$$\langle H_i\rangle_{vacuum} = F_i\; I \neq o\ . \tag{14}$$

The local gauge invariance is then spontaneously broken. All particles become massive so that infrared divergencies disappear.

The effect of the operator A(C) in this model is easily understood when we expand the exponent. The expansion converges by assumption and we find that the operator creates gauge quanta along the curve C. The vacuum expectation value of A(C) is obtained through a Feynman diagram technique where photon propagators are attached both ends onto the curve C. One finds, when C becomes large, that

$$\langle A(C)\rangle_{vacuum} \to \alpha_1\; \exp(-\alpha_2\; L(C)) \tag{15}$$

where L(C) is the total length of C.

The effect of B(C) is entirely different. The model not only contains gauge quanta and Higgs particles but also a locally stable vortex structure, called Nielsen-Olesen vortex or string[7]. The stability of this string is guaranteed by the boundary condition: far away from the vortex we require

$$A_\mu \to \frac{1}{gi}\,\Omega\,\partial_\mu\,\Omega^{-1} \tag{16}$$

where Ω is multivalued when we follow it around the vortex. We observe that B(C) creates just this new boundary condition, so a Nielsen-Olesen vortex is created that coincides with the curve C. Now the physically stable Nielsen-Olesen vortex has a finite width. What B(C) creates is an "excited state" of this vortex that carries more energy per unit of length but whose transverse dimensions start off very small. It will soon decay into the stable vortex plus some gauge quanta. The stable vortex, when created along a closed contour, will finally also decay, but only by shrinking the complete contour defined by its transverse center of gravity. One can show that therefore for large closed curves C,

$$\langle B(C)\rangle_{\text{vacuum}} \to \beta_1 \exp - \beta_2\, \Sigma(C) \tag{17}$$

where $\Sigma(C)$ is the total area spanned by C.
Summarizing: because the gauge quanta are massive the potential between two electric charges decays exponentially with distance, but because the magnetic vortex carries energy per unit of length, the potential between two elementary magnetic charges rises linearly with distance.

A PHASE TRANSITION. THE VACUUM AS SUPER-INSULATOR FOR QUARKS

The way the operators A and B behave in the well-known Higgs model is opposite to what we expect in a quark theory. There it is the *electric* potential which we expect to rise linearly with distance. We now argue that, because the operators A and B are so similar in nature, a phase transition is possible that interchanges the long-distance characters of A and B, if we vary the parameters in the theory. In fact we expect this other phase to be realized also in a model without Higgs fields. What can be proven[5] is the following. Because of the commutation rules (12) the possible ways in which the observables A and B behave for large curves C, C' are restricted. One can have either

Higgs mode:

$$\begin{cases} \langle A(C)\rangle \to \alpha_1 \exp(-\alpha_2\, L(C)) \\ \langle B(C)\rangle \to \beta_1 \exp(-\beta_2\, \Sigma(C)) \end{cases}$$

or
Confinement mode:

$$\begin{cases} \langle A(C)\rangle \to \alpha_1 \exp(-\alpha_2\, \Sigma(C)) \\ \langle B(C)\rangle \to \beta_1 \exp(-\beta_2\, L(C)) \end{cases}$$

or, but unlikely, both Higgs and confinement:

$$\begin{cases} \langle A(C)\rangle \to \alpha_1 \exp(-\alpha_2\, \Sigma(C)) \\ \langle B(C)\rangle \to \beta_1 \exp(-\beta_2\, \Sigma(C)) \end{cases}$$

or an infrared divergent mode with massless particles, presumably representing the critical point where the phase-transition occurs. One cannot have

$$\begin{cases} \langle A(C)\rangle \to \alpha_1 \exp(-\alpha_2\, L(C)) \\ \langle B(C)\rangle \to \beta_1 \exp(-\beta_2\, L(C)) \end{cases}$$

corresponding to a state with neither Nielsen-Olesen vortex nor quark confining strings.

The Higgs model can be regarded as a non-Abelian generalization of a super conductor with the Nielsen-Olesen vortex describing the Meissner effect.

The confinement mode is the dually opposite of a super conductor. The vacuum is a super insulator for quarks.

Since the flux conservation law only holds modulo N, the model will allow N quarks to sit together with finite total energy. Baryons are observed to be composed of 3 quarks. Apparently for them N = 3.

OPEN QUESTIONS AND PROBLEMS

The theory sketched above can be worked out much more quantitatively but still there are many problems to be solved. First, we did not prove that quark confinement occurs, only that it is one of the few possible modes that are separated by phase transition points. It is extremely hard, in fact not yet possible, to do reliable calculations in a pure gauge theory with fermions. We do not know how to compute baryon and meson mass spectra reliably, let alone their cross sections. We believe that the Lagrangean that is written down for the model does indeed describe everything (apart from a by now well understood vacuum degeneracy[8]) but we have no mathematical proof of this, so it could be wrong.

Secondly, the model does have some (though few) degrees of freedom. There is a free parameter g, fixing the mass scale, and then there is a free parameter corresponding to the "mass" of each quark multiplet added. The total number of quark multiplets is free as long as it does not exceed 16. At present the number is 5. If it does exceed 16 then definitely new physics must come in at very small distances.

Are these parameters really free? is there no closer link between strong and weak interactions? We do not know. There are many speculalations on unification of strong and weak interactions, but these give so many possibilities that the answer might have to depend on either super energetic accelerators or fundamentally new theoretical ideas.

REFERENCES

1. C.N. Yang and R.L. Mills, Phys. Rev. 96, 191 (1954).
2. R. Shaw, Cambridge Ph.D. Thesis, unpublished.
3. J. Kogut and L. Susskind, Phys. Rev. D9, 3501 (1974).
   K.G. Wilson, Phys. Rev. D10, 2445 (1974).
4. H.D. Politzer, Phys. Rev. Lett. 30, 1346 (1973).
   D.J. Gross and F. Wilczek, Phys. Rev. Lett. 30, 1343 (1973).
5. G. 't Hooft, Utrecht preprint, to be published in Nucl. Phys. B.
6. F. Englert and R. Brout, Phys. Rev. Lett. 13, 321 (1964).
   P.W. Higgs, Phys. Lett. 12, 132 (1964), Phys. Rev. Lett. 13, 508 (1964), Phys. Rev. 145, 1156 (1966).
7. H.B. Nielsen and P. Olesen, Nucl. Phys. B61, 45 (1973).
   B. Zumino, in Renormalization and Invariance in Quantum field theory, ed. by E.R. Caianiello (1974 Plenum Press, New York).
   Y. Nambu, Phys. Rev. D10, 4262 (1974).
8. S. Coleman, Lectures delivered at the 1977 Int. School of Subnuclear Physics, Ettore Majorana.

QUARK CONFINEMENT AND LATTICE CALCULATION*

Kenneth G. Wilson
Laboratory of Nuclear Studies, Cornell University, Ithaca, NY 14853

ABSTRACT

Quark confinement is characterized as being a consequence of color fields with lines of force restricted to tubes of a fixed transverse size. The restriction is argued to be valid if the color field has quanta with nonzero mass. A generalized loop integral is proposed as the best signal to use in numerical calculations looking for confinement. Monte Carlo methods are proposed for numerical calculations which do not require perturbation theory.

Professor Dirac was interested from the beginning of his career in approximation methods. The Monte Carlo methods, discussed at the end of this paper, are perhaps not as elegant as the methods Professor Dirac favors, but I hope that they will nevertheless provide new insights into the quantum field theories that Professor Dirac helped to invent.

The problem of quark confinement is a basic question at present for the theory of quantum fields and for quantum chromodynamics in particular. In the previous paper, 't Hooft has discussed the problem of characterizing a confined quark phase, if it exists. The lattice gauge theory has been used in the past to help understand the properties of a confined quark theory. But it is important to go beyond the characterization of a confined quark theory and determine whether confinement actually occurs for quantum chromodynamics.

The author plans to use numerical Monte Carlo calculations combined with lattice gauge theory and renormalization group ideas to try to determine whether the pure SU(2) and SU(3) color gauge theories confine quarks. The calculations will be limited (at least initially) to the pure gauge theories without quark fields because the author does not know of any Monte Carlo methods for integration over Dirac fields. The Monte Carlo calculations will require prodigious amounts of computer time; the author owns a share in an Array Processor which will provide enough computing power to start the calculations.

In this paper the background for the Monte Carlo calculations will be reviewed. First the quark-string picture of confinement will be described, with the string being realized in terms of the lines of force of the color gauge field. This picture has been emphasized by Susskind, et al.[1] ; it is related also to the magnetic vortex line pictures of Nielsen and Olesen[2] . Secondly, the characterization of confinement in the pure gauge theory will be discussed. A generalization of the usual gauge

*Supported in part by National Science Foundation grant.

ISSN: 0094-243X/78/009/$1.50

field loop integral will be introduced, which is easier to work with numerically. The generalization involves a sum over many loops. The behavior of the generalized loop integral versus the size of the loop will be used to signal the presence or absence of confinement. Finally the general structure of a numerical calculation will be explained with some estimates of the computing time required to begin the calculations.

In the lines-of-force picture of confinement, there are lines of force between any pair of quarks representing the color field generated by the quarks. Over short distances these lines of force spread out as for the classical Coulomb field. The energy in the field decreases as the lines of force thin out, and as a result the potential energy of the quarks decreases with separation as 1/r, the Coulomb form. But for large distances where confinement occurs the lines of force do not spread any more. Instead, as shown in Fig. 1, they occupy a sausage like region of fixed transverse size. In the middle region of the sausage the density of lines of force is independent of the length of the sausage. This means there is a constant energy density per unit length in the sausage. Therefore, the total energy in the color field is proportional to the distance between the quarks, leading to the now familiar linear potential.

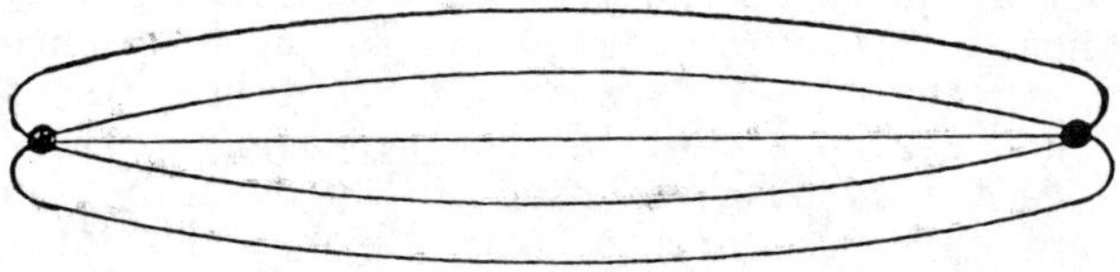

Fig. 1 Lines of force between quarks confined to a tube.

This is the picture of confinement using lines of force. The question now is, why would the lines of force be confined to a tube instead of being spread out over all space? We are clearly interested in the minimum energy that a state with two well separated quarks can have, and the Coulomb distribution gives a lower energy than the sausage form. Thus, it seem unlikely that the sausage distribution occurs in practice.

There is a fallacy in the above estimate of the energy of the Coulomb distribution of lines of force. Namely, the energy estimate is a classical estimate ignoring quantum fluctuations. If the quantum fluctuations in the field are large, the energy of a particular field configuration may be determined by the size of the fluctuations about the field configuration instead of the size of the average field. In this case there is no benefit to having the average (classical) field being small. In quantum field theory the importance of quantum fluctuations increases as the coupling strength increases. In particular, quantum fluctuations have a large effect in quantum chromodynamics at long distances because the effective coupling strength is large for large distances. Thus fluctuations are more important for the quark color field than for electromagnetic fields.

For a correct energy calculation the lines of force must be discussed quantum mechanically. This means we need a quantum mechanical definition of the lines of force, and procedures for estimating their effects.

Physical quantities are represented quantum mechanically by operators. An operator representing a line of force is easily defined. Consider the operator

$$P \exp \{i\, g_o \int_L A^a_\mu(y)\, dy^\mu .\, T^a\} \quad (1)$$

where $A^a_\mu(y)$ is the color gauge field (a is the color octet index, present because the gauge field is an octet under color transformations). The matrices T^a are the 3x3 matrix generators of color SU(3). The integration path L is the path of the line of force. In the simplest case the path of the line of force runs from the position of a quark to the position of an antiquark. g_o is the coupling constant. P is a path ordering symbol, required because of the non-commuting matrix generators.

Exponentials of the line integral of a gauge field have been used in the past to construct gauge invariant products of charged fields. In the case of quantum chromodynamics, the product

$$\psi(0)\, P \exp \{\, i\, g_o \int_L A^a_\mu(y)\, dy^\mu .\, T^a\}\, \overline{\psi}(x) \quad (2)$$

(where $\psi(x)$ is the quark field) is gauge invariant provided the path L runs from the origin to the point x. It is not obvious that the exponentials of line integrals have any connection to the classical lines of force. This does not matter; what is important is that these operators can be used to discuss the minimum energy state of a well separated quark-antiquark pair. In the discussion several analogies will appear between the line of force operator and classical lines of force.

In classical physics the lines of force can begin and end only at a source, such as a charged particle. If charged particles are present, there must be lines of force which begin or end at each particle. The quantum mechanical analogy to this situation comes from the requirement that any physically observable quark-antiquark state must be gauge invariant. The simple product $\psi(0)\overline{\psi}(x)$ is not gauge invariant under color gauge transformations; thus we do not want to use $\psi(0)\overline{\psi}(x)$ as the operator creating a quark-antiquark pair. The operator (2) is gauge invariant. However, this operator corresponds to only a single line of force. To get a distribution of lines of force one can form a sum over many such operators, using a different path L for each member of the sum. To create a quark-antiquark state, this operator is applied to the vacuum state:

$$\int_L \rho(L)\, \psi(0)\, P \exp\{i\, g_o \int_L A^a_\mu(y)\, dy^\mu\, T^a\}\, \overline{\psi}(x)|\Omega> \qquad (3)$$

where $\rho(L)$ is a density function for lines of force and $\int_L$ is an integral over the possible paths from 0 to X.

The next step is to find the distribution of lines of force in expression (3) that gives the minimum energy quark-antiquark state. This problem will be discussed very qualitatively. The immediate aim will be to define circumstances in which the string-like sausage distribution gives the lowest energy. First we must make sure that the Coulomb distribution is obtained when a 1/r field-energy is expected.

In quantum field theory, operators create particles. In the familiar weak coupling context, the color gauge fields create zero mass particles called gluons. A single line of force operator creates gluons in the vicinity of the line of force. Since a single line of force has no transverse size, the gluons created mostly are highly localized in the transverse direction which means they have a high energy. A smooth distribution of lines of force creates lower energy gluons and hence lower energy. In particular, for quarks separated by a distance r, the lines of force can spread to a transverse size of r also; then a typical gluon created has momentum $\sim 1/r$ (by the uncertainty principle; h is set to 1) and hence energy 1/r.

The lines of force are necessarily more compact near an individual quark, causing higher energy gluons to be created. However, these gluons contribute to the self-energy of an individual quark and not to the potential energy between two quarks. Thus the potential energy between quarks behaves as 1/r.

In weak coupling there is a further reduction in the potential energy due to the fact that mostly the exponential of the gauge field is 1. Only in order g_o does the exponential involve A^a_μ and create gluons. Since the gluon state occurs with amplitude g_o, it occurs with probability of order g_o^2. Hence the potential energy is g_o^2/r.

The alternative to zero mass quanta for the gauge field is to have finite mass quanta. This is possible only in strong coupling; for weak coupling we know that gluons have zero mass from the explicit perturbative solution of the gauge field theory. But for long distances the effective color gauge field coupling strength is large and the quantum fluctuations make it difficult to obtain explicit properties of the gauge field. In the absence of any more precise information, the existence of non-zero mass quanta has to be considered as one option the theory might take.

The characteristic feature of a quantum field with non-zero mass quanta is that there are no correlations in the field over distances longer than the Compton wavelength of the quanta. As a result, with finite mass quanta the lines of force in different regions of space separated by more than a Compton wavelength, act independently. In particular, the region between the quark and antiquark can be broken into blocks the size of a Compton wavelength. The lines of force going through these regions act now like a product of operators for each region separately (since the line of force operator is an exponential, which is a product). Each operator in the product creates a quantum, whose energy is roughly its mass m. The minimum energy is now obtained by minimizing the number of blocks between the quark and antiquark. The resulting blocks combine to form the sausage of Fig. 1. Any other path between the quark and antiquark is longer, contains more blocks, and has higher energy. Hence the lowest energy distribution for the lines of force is the sausage distribution.

The question of how confinement occurs has not been answered by the above discussion; all it does is transpose the question into another question: do the quanta of the color field have a nonzero mass?

I have no arguments to answer this question. However, a calculation can be defined that may help to answer this question. A non-perturbative calculation is required since in perturbation theory the mass is zero. Thus, the proposed use of Monte Carlo methods (see later).

The simplest calculation to formulate is the expectation value of a single closed line of force operator. The line of force typically has the form of a square of size L with one side being timelike, the other direction being spacelike. This expectation value is calculated in a Euclidean metric (imaginary time). The interpretation of this expectation value is that it represents the expectation value of e^{-HL} for a heavy quark-antiquark state connected by a single spacelike line of force. The time-like sides of the loop are necessary to describe the time propagation of heavy quarks. The Euclidean metric means the time propagation operator is e^{-Ht} not e^{-iHt}. Since the two spacelike sides of the square are separated by time L the expectation value is of e^{-HL}. Typically the expectation value is calculated for the pure gauge field, so only the gauge field energy is included in H. If L is large, the operator e^{-HL} projects out the minimum energy component of the state it is applied to and hence its expectation value is of order $e^{-E_{min}L}$ where E_{min} is the minimum gauge field energy of the quark antiquark pair.

In weak coupling the principal energy in E_{min} is the self energy of each single quark caused by the gauge field. Since this energy is independent of L, the expectation value of the closed line of force behaves as e^{-cL} where c is a constant energy independent of L.

For the nonzero mass case the energy behaves as L, namely as m(mL) where m is the mass of the gauge field quantum and (mL) is the number of Compton wavelengths separating the quarks. Hence the expectation value behaves as $e^{-m^2L^2}$, or the exponential of the area enclosed by the loop.

Unfortunately, it is difficult to distinguish e^{-cL} from $e^{-m^2L^2}$ when L is large. In any kind of rough calculation, both expressions are indistinguishable from 0.

The e^{-cL} behavior in weak coupling is caused by the short-distance contribution of the gluon field to a single quark energy. This is of no interest. To eliminate it one can propagate an extended distribution of charge, in place of a point charge. This means finding the expectation value of an extended distribution of closed lines of force, instead of a single line of force. By using a distribution of width about L, when the overall size of the loop is L, one has a system whose total energy should come from low energy gluons (in the weak coupling case); their energy should be of order 1/L so the expectation value of e^{-HL} is of order 1. (More precisely, it is equal to 1 up to terms of order g_0^2 for weak coupling. One expects this expectation value, therefore, to behave as $e^{-g^2(L)}$ where g^2 (L) is the effective coupling constant for scale L, at least for weak coupling.) In the limit of large L we expect the expectation value to go to a constant if the quanta have zero mass. In contrast, for nonzero mass, the energy should be linear in L, as argued before, and therefore the expectation value should go to 0 as $e^{-m^2L^2}$. Thus using the expectation value of a broad distribution of closed lines of force gives a sharper test of the two alternatives.

The final topic of this talk is the application of Monte Carlo methods to compute expectation values such as for a distribution of closed line of force operators. The particular variant of the Monte Carlo method that is useful for this problem is the method of Metropolis et al[3], which has been widely used in statistical mechanics. The Metropolis method is used to compute the ratio of two integrals. This is appropriate for the calculation of expectation values in field theory, which have the form

$$\langle F \rangle = Z^{-1} \int_{A_\mu} F(A_\mu)\, e^{A[A_\mu]} \tag{4}$$

with

$$Z = \int_{A_\mu} e^{A[A_\mu]} \tag{5}$$

where $\int_{A_\mu}$ refers to the functional integral over all gauge field functions $A_\mu^a(x)$; $A[A_\mu]$ is the action and $F(A_\mu)$ is the function whose expectation value is desired.

The Metropolis method involves cycling through the integration variables, one by one. This is clearly not possible for a functional integral, for which the integration variables cannot be enumerated. As a result the continuum gauge theory has to be approximated or replaced by the lattice gauge theory, which has a discrete set of field variables to be integrated over. In the lattice gauge theory there is a hypercubic lattice of points in spacetime, with a lattice spacing a that can be chosen arbitrarily. The basic fields are short line of force operators connecting adjacent lattice sites. These operators, classically, are group elements of the SU(3) group. Integration over these elements consists of invariant group integration.

Using the lattice theory one has an invariant SU(3) group integration to perform for each nearest neighbor link. The group element for the link from lattice site in the direction μ will be denoted $U_{n\mu}$. The Metropolis method is a procedure for generating a sample of sets $\{U_{n\mu}\}$(from a finite subset of the lattice) where the distribution function of the sample is given by the integrand exp(A) of Eq. (5). Then the normalized probability distribution for the sample is e^A/Z; as a result, the expectation value of F can be estimated by computing the average of F for the samples generated.

The Metropolis procedure, adapted to group integration, proceeds as follows. One chooses, arbitrarily, the initial sample $\{U_{n\mu}^{(0)}\}$. To generate the next sample one cycles through all the links in turn. For a given link $n\mu$, the procedure is, first, to generate, tentatively, an updated $U_{n\mu}$. This is done by multiplying the old matrix $U_{n\mu}$ by a random SU(3) matrix V. The easiest procedure for generating V is to have a table of matrices V, say 16 of them, and choose randomly one of the 16. The only restriction on the table is that if a matrix V appears in the table, then the conjugate matrix V^+ must also appear. It is also necessary that the set of all products of arbitrary order of members of the table must be everywhere dense in the group space. The list need not be uniformly distributed in group space; in fact, for weak coupling the matrices on the list should all be close to the unit matrix.

The second part of the procedure is to decide whether to accept the new matrix or to keep the old matrix in the new sample. To decide this the ratio of the integrands e^A is computed for the new $U_{n\mu}$ versus the old $U_{n\mu}$. If the integrand is increased by using the new $U_{n\mu}$, then the new $U_{n\mu}$ is accepted. If the integrand is decreased, then a random number between 0 and 1 is selected. If the ratio of new to old integrand is greater than the random number, the new $U_{n\mu}$ is accepted. Otherwise, the old value is kept.

Once a cycle through all the link variables is complete the cycle is repeated; to produce a reasonable sample size the whole cycle has to be repeated roughly 10000 times. (10000 is a preliminary guess since the calculations have not been performed yet).

The proof that the probability distribution for the samples is e^A/Z is fairly straightforward and will not be repeated here.

A reasonable size lattice for an actual calculation might contain 8 lattice sites on a side. Such a lattice, with periodic boundary conditions, contains 16384 links. Thus a Monte Carlo calculation of 10000 cycles on this lattice involves about 160 million upgrading steps. The author has a share of an array processor (made by Floating Point Systems, Portland, Oregon) which completes one upgrade in about 160 microseconds for the SU(2) gauge theory (this is being studied first to save computer time). For comparison, on the IBM 370-168 the same calculation takes about 600 microseconds. The whole calculation requires about 7 hours of Array Processor time. To do a thorough study of the gauge theory will require prodigious amounts of computer time. Fortunately, the cost of computing on the array processor is roughly 60 times less than the IBM 370-168 which makes the calculations feasible.

REFERENCES

1. See, e.g., A. Casher and L. Susskind, Phys. Rev. D9, 436 (1974).
2. H. Nielsen and P. Olesen, Nucl. Phys. B61, 45 (1973).
3. See, e.g., J. M. Hammersley and D. C. Handscomb, Monte Carlo Methods, (Wiley, New York, 1964).

A REVIEW OF DEVELOPMENTS IN LATTICE GAUGE THEORY: EXTREME ENVIRONMENTS AND DUALITY TRANSFORMATIONS

John B. Kogut†
Cornell University, Ithaca, N.Y. 14853

ABSTRACT

Recent calculations showing that various lattice gauge theories lose their quark confining property abruptly at a finite, large temperature are reviewed. Models of confinement at finite baryon density are discussed. Duality transformations and the phase diagrams of Abelian gauge theories in ordinary environments are reviewed. The use of exact and approximate duality transformations on gauge theories in general is stressed as a useful new tool.

INTRODUCTION

I would like to review some recent work on lattice gauge theories. First, we discuss the possibility that these theories lose their quark confining property when placed in extreme environments such as high temperature[1,2] or density.[3] The second topic reviews duality transformations of Abelian and Z_N lattice gauge theories which relate them to more familiar physical systems whose qualitative properties are known. For example, Abelian lattice gauge theories can be mapped onto the theory of interacting closed threads[4] which is in turn related (approximately) to conventional scalar electrodynamics.[5]

The first topic, confinement in extreme environments, may have astrophysical implications. The large baryon density in some neutron stars and the high temperatures present in the early stages of the universe may be sufficiently extreme environments so that strongly interacting matter is better described as quarkium than as nuclear matter.[6] Recent work on lattice gauge theories indicate that as the temperature T is increased, a critical point T_c is reached where confinement is abruptly lost.[1,2] In an Abelian lattice theory with no quarks the potential energy V(R) of an external static quark antiquark pair whose members are a distance R apart changes from linear,

$$V(R) = \text{const. } |R|, \qquad T < T_c \tag{1a}$$

to Coulombic,

$$V(R) = Q^2/|R|, \qquad T > T_c \tag{1b}$$

above the critical point. The phase transition at T_c is second order. An even more interesting phenomenon occurs in non-Abelian lattice gauge theories of pure gluonic matter: below T_c Eq. (1a)

†A. P. Sloan Foundation Fellow. Supported in part by the National Science Foundation.

ISSN: 0094-243X/78/017/$1.50

applies, but above the transition the force law become short-ranged,

$$V(R) \sim e^{-\mu|R|}, \qquad T > T_c \tag{2}$$

How is it possible that quarks which are in the fundamental representation of the gauge group can be screened by gluons which lie in the octet representation? This behavior is analogous to placing a charge of arbitrary strength in a plasma and finding that there is no residue of Coulomb's law. Apparently the charge fluctuations of the medium are sufficient to screen any impurity, even its charge is incommensurate with the fundamental charges in the medium. So, Eq. (2) teaches us that the non-Abelian gluon medium forms a non-Abelian plasma which can screen quarks of any color.

One would guess that the critical temperature is on the order of $kT_c \approx 1$ GeV, a typical hadron mass. If so, this phase transition might have only been relevant in the earliest stages of the creation of the universe. Extreme environments of a less remote variety may occur in the interiors of neutron stars--here the baryon density is suspected to be several times that of ordinary nuclear matter. In this environment the nucleons are overlapping considerably so one is led to suspect that quarks are more relevant to a description of the system than protons and neutrons. To address this question one would like to place QCD into an environment of variable baryon density b and search for qualitative changes in the theory's character as b ranges from zero upward. Such a study has been made in 1+1 dimensional models of confinement and a rich spectrum of phenomena have been found.[3] The $q\bar{q}$ potential weakens continuously as b increases in some of these models, i.e. without an abrupt phase transition. Systematic studies of lattice gauge theories in 3+1 dimensions in the presence of a background baryon density have not been carried out.

In this talk we emphasize lattice gauge theories and some models of confinement. But the questions posed here have been studied in continuum QCD as well. It has been argued that at high temperature and/or high density many properties of these theories can be analyzed using renormalization-group-improved perturbation theory. The reason for this is, roughly, that high temperature or density bring a large momentum scale into the problem (through large kinetic energies $\sim$ kT, or a huge Fermi energy) about which renormalization-group-improved perturbation theory can be performed. Since these theories are asymptotically free such calculations are reliable. Among the results obtained in this way is the computation of the equation of state (free quarks plus calculable corrections) for QCD in a large baryon density and a derivation of the plasma phase of QCD at high temperature. The lattice and model calculations add more detail to these results, e.g. the behavior of the theories at all temperatures and densities, precise predictions in the vicinity of phase transitions, etc. In ref. 7 I have attempted to collect a partial list of contributions to the continuum QCD studies. (It is not possible to compile a complete list--I apologize in advance to any author whose work I have carelessly omitted.)

Before discussing these topics in detail, one should realize that the results cited above mean that QCD in extreme environments becomes rather conventional. In ordinary environments QCD presumably confines the quanta created by its fundamental fields. However, at high temperature, for example, this will not be the case. This fact undermines a frequent criticism of QCD which objects to any theory whose fundamental quanta cannot be isolated. The advocates of this view claim that "if the fundamental quanta cannot be isolated then they must be irrelevant .... a more direct, simpler version of strong interactions should be formulated without the excess baggage of quarks and gluons." Since QCD liberates its quarks and gluons at high temperatures, these quanta are manifestly "real" and this objection loses much of its impact.

EXTREME ENVIRONMENTS

Let's consider the argument that Abelian lattice gauge theory undergoes a second order phase transition at finite temperature from a quark confining to an ordinary theory.[1,2] The strategy of the analysis consists in relating this model to a simpler, more familiar spin lattice whose phases are known. Then the calculation of the $q\bar{q}$ potential is mapped into the calculation of the spin-spin correlation function and the results of Eq. (1) and (2) are read off. The correspondences are:

$$\begin{aligned} &\text{Lattice Gauge Theory Partition Function} &&\Leftrightarrow\ \text{3-dimensional XY Model Partition Function} \\ &\text{Temperature} &&\Leftrightarrow\ \text{Inverse Temperature} \\ &q\bar{q}\ \text{Potential} &&\Leftrightarrow\ -\ell n[\text{Spin-Spin Correlation Function}] \end{aligned} \tag{3}$$

These correspondences establish a duality relation between the two models. Since it is known that the 3-dimensional XY model has a second order phase transition at T*, the correspondences imply the existence of a phase transition in the Abelian lattice gauge theory at a temperature T_c. At a temperature in the XY model above T*, the spin-spin correlation function $\langle s(R)s(0)\rangle$ decays exponentially with R. Using the correspondences, this means that the lattice gauge theory confines for $T < T_c$,

$$\begin{aligned} V(R) &\sim -\ell n \langle s(R)s(0)\rangle \\ &\sim -\ell n\, e^{-\mu|R|} \\ &\sim \mu|R| \end{aligned} \tag{4}$$

Similarly, at a temperature below T* in the XY model the spin-spin correlation function approaches a constant, the magnetization squared, at a rate determined by spin wave analysis, $\langle s(R)s(0)\rangle \sim m^2 e^{-c/|R|}$, as $R \to \infty$. This means that for $T > T_c$ in the lattice gauge theory,

$$V(R) \sim -\ln \langle s(R)s(0)\rangle$$
$$\sim -\ln(m\, e^{-c/|R|})$$
$$\sim c/|R| + \text{const.} \tag{5}$$

which reproduces Coulomb's law!

Let's establish the first part of the duality relations. The Partition Function for Abelian lattice gauge theory at finite $\beta = 1/kT$ is

$$Z(\beta) = \sum_{\substack{\text{physical}\\ \text{states}}} e^{-\beta\cdot(\text{energy of state})} = \operatorname*{Tr}_{\substack{\text{physical}\\ \text{states}}} e^{-\beta H} \tag{6}$$

where H is the Hamiltonian of the theory (time is continuous and there are 3 discrete spatial axes: The lattice spacing is "a"). The restriction of the sum in Eq. (6) to "physical states" is important. To appreciate it recall the construction of the Hamiltonian form of the theory.[8] Space is discrete with sites labelled r, a triplet of integers, and directed links $(r,\hat{n})$, which begin on site r and point one lattice unit in the $\hat{n}$ direction. The degrees of freedom of the theory consist of phases

$$e^{i\phi(r,\hat{n})} \tag{7}$$

defined on links, and there are variables $E(r,\hat{n})$ conjugate to the angular variables $\phi(r,\hat{n})$,

$$[\phi(r,\hat{n}), E(r',\hat{n}')] = i\delta_{r,r'}\delta_{n,n'} \tag{8}$$

(ϕ is related to the vector potential $\underset{\sim}{A}$ of QED, $\phi(r,\hat{n}) \to g\underset{\sim}{A}(r)\cdot\hat{n}a$, and E is the electric flux operator, $E(r,\hat{n}) \to \underset{\sim}{E}(r)\cdot\hat{n}a^2$. One can check that Eq. (8) reproduces the canonical commutation relations between $\underset{\sim}{A}$ and the electric field $\underset{\sim}{E}$ of QED formulated in the $A_0 = 0$ gauge when the continuum limit $a \to 0$ is taken.)

The Hamiltonian of the lattice theory consists of two pieces: a lattice analog of the electric field squared, and a lattice analog of the magnetic field squared. The first term is easily constructed since $E(r,\hat{n})$ is the operator which measures the electric flux passing from site to site,

$$\frac{1}{2}\int \underset{\sim}{E}^2\, d\underset{\sim}{x} \to \frac{g^2}{2a} \sum_{\text{links}} E^2(r,\hat{n}) \tag{9}$$

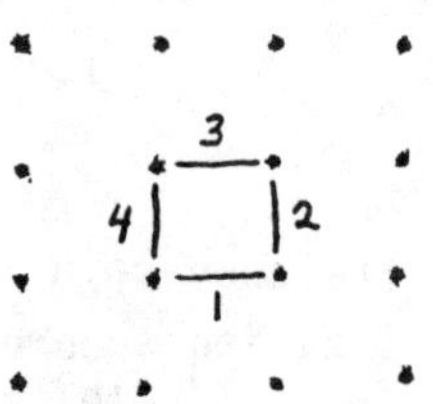

Fig. 1
A plaquette

An interesting lattice form of the square of the magnetic field makes use of the phase variables introduced in Eq. (7). Consider four links 1,2, 3, and 4 which form a closed square, Fig. 1. Associate with each square the product of phases,

$$\exp i(\phi(1) + \phi(2) + \phi(3) + \phi(4))$$

Then a possible correspondence is[9,10]

$$\frac{1}{2}\int \underset{\sim}{B}^2 \, d\underset{\sim}{x} \rightarrow -\frac{2}{ag^2} \sum_{\text{squares}} \cos(\phi(1) + \phi(2) + \phi(3) + \phi(4)) \quad (10)$$

(One can check that this correspondence becomes an equality in the classical continuum limit. To do this, observe that Stokes' law and the relation $\phi(r,\hat{n}) \rightarrow g\underset{\sim}{A}(r)\cdot\hat{n}a$ imply that the sum of ϕ's around a square is proportional to the magnetic flux passing through the square, and that if ϕ is a smooth field and the $a \rightarrow 0$ limit is taken only the quadratic term in the cos term survives in Eq. (10).) The lattice Hamiltonian now reads[8]

$$H = \frac{g^2}{2a} \sum_{\text{links}} E^2(r,\hat{n}) - \frac{2}{ag^2} \sum_{\text{squares}} \cos(\phi(1) + \ldots + \phi(4)) \quad (11)$$

The important features of this expression are that it has exact gauge invariance for any lattice spacing, and that ϕ is effectively bounded, $-\pi < \phi < \pi$.

This formulation of the theory is fully specified once the physical space of states is defined. Recall that the quantization of electromagnetism in the class of gauges $A_0 = 0$ requires a subsidiary condition. It reads

$$\sum_{\hat{n}} E(r,\hat{n})|\text{physical state}\rangle = 0 \quad (12)$$

which is a discrete form of Gauss' law. This constraint and the gauge condition reduce the number of independent degrees of freedom of the gauge field down to two. Eq. (12) allows us to picture physical states as closed loops of electric flux constructed so that electric flux is conserved at each site. Furthermore, since $E(r,\hat{n})$ is conjugate to an angular variable $\phi(r,\hat{n})$ the spectrum of $E(r,\hat{n})$ consists of just the integers--electric flux is quantized in this theory. So, the flux a link can be 0, ±1, ±2, etc. This allows the physical states to be enumerated in a countable fashion. Some examples are shown in Fig. 2.

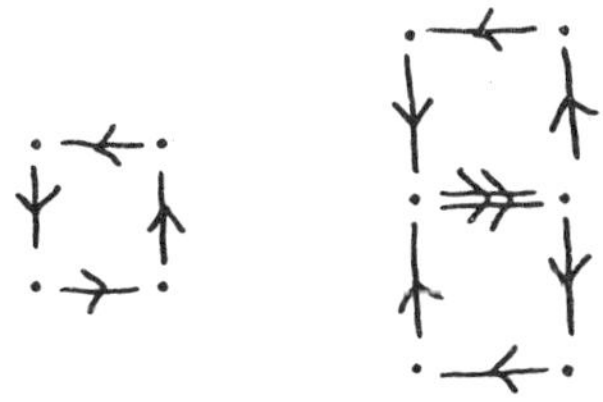

Fig. 2 Physical states

Now we can turn back to the expression for the Partition function, Eq. (6). The sum over physical states can be replaced by a constrained sum over all states

$$Z(\beta) = \sum_{\substack{\text{all}\\ \text{states}}} \prod_r \delta_{\nabla\cdot E(r),0} e^{-\beta(\text{energy of state})} \quad (13)$$

The Kronecker symbols enforce the constraints of Eq. (12). Now let's simplify the problem of computing $Z(\beta)$: replace the Hamiltonian by just the electric flux term,

$$H \rightarrow \frac{g^2}{2a} \sum_{\text{links}} E^2(r,\hat{n}) \quad (14)$$

This replacement is justified because the electric term in H guarantees confinement at T = 0. Using Eq. (14) we shall still find that the theory loses confinement at a finite T_c. Therefore, the inclusion of the magnetic term, which weakens the confining potential at T = 0, can only reduce T_c--it could not change the argument that a finite T_c exists. Recall how the electric term in H leads to confinement. Place a static quark at position r = 0 on the lattice and a static antiquark at position r = R. To be physical this state must satisfy Gauss' law. This means that a unit of electric flux must connect the $q\bar{q}$ pair. The lowest energy configuration of flux is given by the shortest path between the pair, Fig. 3. But according to Eq. (14) each link on this path costs an energy $g^2/2a$ and there are R/a links, so the total energy is $g^2/2a^2 \cdot |R|$--the linear confining potential.

Fig. 3 Static quarks and flux

At this stage the Partition function is

$$Z(\beta) = \sum_{\substack{\text{all}\\ \text{states}}} \prod_r \delta_{\nabla\cdot E(r),0} \exp\left(-\beta \frac{g^2}{2a} \sum_{\text{links}} E^2(r,\hat{n})\right) \tag{15}$$

This is the Partition function for a three-dimensional system of closed ($\nabla\cdot E = 0$) threads. The aficionado of spin systems will recognize Eq. (15) as the partition function of the XY model in three dimensions[4,11]--the correspondences in Eq. (3) are immediate to him and he is done. Let's work out these connections.

Consider a three-dimensional lattice whose sites are occupied by two dimensional spins,

$$s(r) = \begin{pmatrix} \cos\theta(r) \\ \sin\theta(r) \end{pmatrix} \tag{16}$$

The Hamiltonian consists of nearest neighbor ferromagnetic couplings. In terms of $\theta(r)$ we have

$$H = J \sum_{r,\mu} (1 - \cos\Delta_\mu\theta(r)) \tag{17}$$

in an abbreviated but suggestive notation in which Δ_μ is a discrete difference operator ($\Delta_\mu f(r) \equiv f(r+\hat{\mu}a) - f(r)$) in the direction μ (μ=1,2,3) between sites. The statistical mechanics of this model follows from the Partition function,

$$Z(\beta) = \prod_r \int_{-\pi}^{\pi} \frac{d\theta(r)}{2\pi} \exp\left[-\beta \sum_{r,\mu} (1 - \cos\Delta_\mu\theta(r))\right] \tag{18}$$

This expression is difficult to work with so we will replace it by a simpler model which has been analyzed in detail. To motivate it, we note that the integrand of $Z(\beta)$ is a periodic function of $\theta(r)$. Therefore, it can be written as a Fourier series. For large β (small temperature) the Fourier transform of $\exp(\cos\Delta_\mu\theta(r))$ is well approximated by a Gaussian,

$$\exp(\beta\cos\Delta_\mu\theta) \rightarrow \sum_{\ell_\mu=-\infty}^{\infty} \exp(+i\ell_\mu\Delta_\mu\theta)\exp(-\ell_\mu^2/4\beta) \tag{19}$$

The right-hand side of this replacement preserves all the important features of the original spin lattice and is numerically precise for low temperature. But the right-hand side is well-defined for all β and is more easily analyzed because of its Gaussian form. It is referred to as the Villain model[12] and has been studied in its own right. We now concentrate on it and show that its partition function is dual to the Abelian lattice gauge theory at finite temperature. Making the replacement Eq. (19) in the Partition function Eq. (18), we see that each $\theta(r)$ integral can be done and that each generates a familiar constraint, $\Delta_\mu\ell_\mu(r) = 0$. Now, Eq. (18) becomes

$$Z = \sum_{\ell_\mu(r)=-\infty}^{\infty} \prod_r \delta_{\Delta_\mu\ell_\mu(r),0} \exp(-1/4\beta \sum_{r,\mu} \ell_\mu^2(r)) \tag{20}$$

Comparing Eq. (15) with Eq. (20) we have established the first part of the duality relation: Z(lattice model) $\rightarrow$ Z(spin system) if T(lattice model) $\rightarrow$ 1/T (spin system). Since the Villain model is known to have a second order phase transition at $T_c \leq 6.2$,[13] we learn that the lattice model also has two phases. We shall now see that the phases can be labelled by different qualitative behaviors of the $q\bar{q}$ potential and that the potential is dual to the spin-spin correlation function of the Villain model.

To calculate the $q\bar{q}$ potential we must compute the Partition function with a source of one unit of flux at $r = 0$ and a sink of one unit of flux R lattice sites away. Now the expression for Z is the same as in Eq. (15) except the constraint at $r = 0$ reads $\delta_{\nabla\cdot E,1}$ and that at $r = R$ reads $\delta_{\nabla\cdot E,-1}$.

$$Z^{q\bar{q}} = \sum_{\substack{\text{all}\\\text{states}}} \prod_r \delta_{\nabla\cdot E(r),Q(r)} \exp\left\{-\beta \frac{g^2}{2a} \sum_{\text{links}} E^2(r,n)\right\} \tag{21}$$

where $Q(r) = \delta_{r,R} - \delta_{r,0}$ is the charge density of the external quarks. The inter-quark potential is the free energy of this system minus the free energy of the theory without the external quarks,

$$V(R) = -\frac{1}{\beta}[\ln Z^{q\bar{q}} - \ln Z] = -\frac{1}{\beta}\ln(Z^{q\bar{q}}/Z) \tag{22}$$

Keeping this result in mind we now turn to the spin-spin correlation function in the Villain model. Consider the function,

$$\begin{aligned} C(R) &= \langle s_1(R) + is_2(R), s_1(0) + is_2(0)\rangle \\ &= \langle e^{i[\theta(R)-\theta(0)]}\rangle \end{aligned} \tag{23}$$

In the spin system Eq. (18) C(R) becomes

$$C(R) = Z(R)/Z \tag{24a}$$

where

$$Z(R) \equiv \prod_r \int_{-\pi}^{\pi} \frac{d\theta(r)}{2\pi} \exp\left\{-\beta \sum_{r,\mu} (1-\cos\Delta_\mu\theta(r)) + i(\theta(R)-\theta(0))\right\} \qquad (24b)$$

Making the Villain replacement, Z(R) becomes more simply

$$Z(R) = \prod_r \int_{-\pi}^{\pi} \frac{d\theta(r)}{2\pi} \sum_{\ell_\mu(r)=-\infty}^{\infty} \exp\left\{-\frac{1}{4\beta} \sum_{r,\mu} \ell_\mu^2(r)\right\} \exp\left\{i \sum_{r,\mu} \ell_\mu(r)\Delta_\mu\theta(r)\right\}$$
$$\times \exp\left\{i[\theta(R)-\theta(0)]\right\} \qquad (25)$$

All the integrations over $\theta(r)$ can be done as before except at the special sites $r = 0$ and $r = R$. The additional factor $\exp(i\theta(R))$ at the site R replaces the constraint by $\Delta_\mu \ell_\mu = 1$, and the factor $\exp(-i\theta(0))$ at the site 0 produces the constraint $\Delta_\mu \ell_\mu = -1$. So, we have

$$Z(R) = \sum_{\ell_\mu(r)=-\infty}^{\infty} \prod_r \delta_{\Delta_\mu \ell_\mu(r),Q(r)} \exp\left\{-\frac{1}{4\beta} \sum_{r,\mu} \ell_\mu^2(r)\right\} \qquad (26)$$

where the "charge" $Q(r) = \delta_{r,R}-\delta_{r,0}$. Thus Z(R) is just the partition for the lattice gauge theory in the presence of a static quark at R and an antiquark at $r = 0$! So, now we have the last line in the Duality relations of Eq. (3),

$$V(R) \sim -\ln C(R) \qquad (27)$$

where the temperature of the spin system maps onto inverse temperature of the lattice gauge theory. The force laws then discussed in the Introduction follow and we are done.

The arguments for the more interesting non-Abelian theory are similar--the gauge theory computations are mapped onto properties of a non-Abelian spin system in an external field. This system's phase diagram is also well known. The results of this analysis have been discussed above.

Now let's turn briefly to environments at ordinary temperatures but large baryon density ρ. Since the Fermi surface is pushed to high energy as ρ increases, one would guess that many features of the theory resemble the theory of free quarks. Is there a phase transition at a critical density ρ_c at which the theory passes from one of confinement to something qualitatively different? This question has not been answered in lattice gauge theory. However, several models of confinement have been studied in detail.[3] One model was the two species massive Schwinger model-QED in 1 time-1 space dimensions with massive fermions. One species is given an (Abelian) color charge +g and the other -g. The spectrum of this theory consists of "hadrons" --colorless bound states of fermions--which interact locally via an interaction density whose strength is proportional to the fermion mass m. An environment which is colorless but rich in baryon number can be constructed and the force law between a static $q\bar{q}$ pair studied in the usual way. For large baryon density ρ, the potential behaves as

$$V(R) \sim \frac{\text{const.}}{\rho} \cdot \sin^2(\pi \frac{e}{g}) \cdot |R| \tag{28}$$

where e is the color charge of the static quarks. The strength of the linear potential falls to zero smoothly as ρ increases indicating the absence of a phase transition. Of course, the equation of state at high ρ is given by that of 2 species of free quarks plus small, calculable corrections. It would be interesting to calculate some dynamical properties of the 2 species Schwinger model at finite baryon density ρ to see the interplay between the "hadron" character and "quark" character of the theory. Of course, one dimensional models of neutron stars may not be particularly good guides to the real world (!), but studies beyond renormalization-group-improved perturbation theory have not been done in QCD and lattice gauge theories have not been considered at high density.

PHASES OF ABELIAN GAUGE THEORIES

The last topic I would like to describe is (unfortunately) also esoteric. Its aim is to understand the phases of Abelian and Z_N lattice gauge theories[14] by relating them to other more familiar models. This work was motivated by the fact that Abelian lattice gauge theory confines for strong coupling. However, for weak coupling it reduces to conventional continuum QED. Therefore, somewhere in the intermediate coupling region a phase transition (or something similar) must occur to separate these two qualitatively different behaviors. Ideally one wants to find an expression for the Partition function of the model which yields a useful physical picture of the critical region. One approach to this problem takes advantage of the recent analyses of Abelian spin systems alluded to in the previous section. In particular, a similar conceptual problem existed a few years ago in the 2 dimensional XY model. At high temperature general theorems assure us the system is disordered and the correlation function of a spin at r = 0 and one at r = R falls exponentially with R. However, high temperature expansions indicated that the theory's susceptibility diverges at a non-zero temperature.[15] In addition, low temperature analyses indicated that spin waves are the only relevant excitations in the system.[12] But spin waves produce a spin-spin correlation function which vanishes as a temperature-dependent power of R. The challenge then was to find the excitations responsible for the qualitative change in the spin-spin correlation function in the critical region. Those excitations turned out

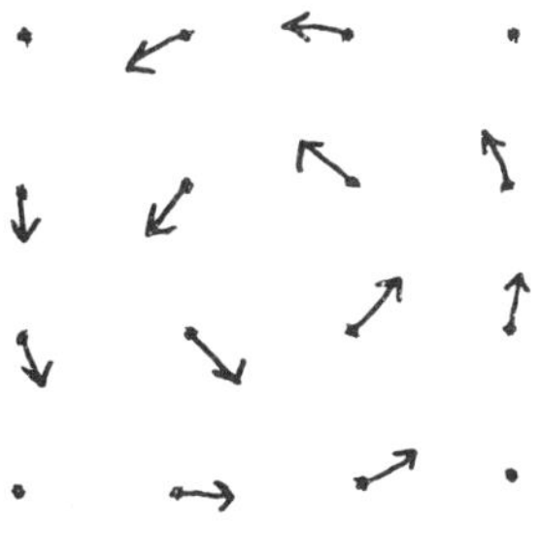

Fig. 4 A vortex

to be vortices--spin configurations which, roughly speaking, have non-zero winding number[16] (Fig. 4). These vortices effect the spins over all of space and clearly tend to disorder the system. A famous energy vs. energy argument[16] indicates that vortices are only important in the statistical mechanics of the system above a critical temperature T_C. Using the Villain approximation and a renormalization group analysis[17], these ideas have been placed on relatively sound footing and an appealing physical picture of the phases of the model has been obtained: For $T < T_C$, only spin wave excitations are important and the spin-spin correlation function is power-behaved. The vortices are bound into small vortex-antivortex pairs and cannot disorder the system. This is the low temperature "dielectric" phase of the model. As T approaches T_C from below the entropy of the system increases until the vortex pairs become unbound (T_C is an ionization point) and the vortices become free. In this "conducting phase" the vortices disorder the system completely.

This bag of tricks can be played on the Abelian lattice gauge theory. Beginning with the space-time symmetric version of the theory[10] (space and time are discrete), the Partition function,

$$Z = \prod_{r,\hat{n}} \int_{-\pi}^{\pi} \frac{d\phi(r,\hat{n})}{2\pi} \exp\left(-\frac{1}{g^2} \sum_{r,\hat{\ell},\hat{n}} [1-\cos(\nabla_{\hat{\ell}}\phi(r,\hat{n})-\Delta_{\hat{n}}\phi(r,\hat{\ell}))]\right) \tag{29a}$$

can be written as[4]

$$Z = \prod_{r,\mu} \sum_{m_\mu(r)=-\infty}^{\infty} \prod_r \delta_{\Delta_\mu m_\mu(r),0} \exp\left(-\frac{\pi}{g^2} m_\mu(r)V(r-r')m_\mu(r')\right) \tag{29b}$$

In Eq. (29a) the unit vector $\hat{n}$ ranges over the <u>four</u> directions of the links in the space-time lattice. The index μ serves the same purpose in Eq. (29b). In that expression V is the four dimensional Coulomb potential. So, Eq. (29b) has changed the problem of understanding the Abelian lattice gauge theory written in terms of the lattice version of $F_{\mu\nu}^2$ into the problem of understanding a four dimensional system of closed current loops interacting through a Coulomb potential, Fig. 5. These loops are analogous to the vortices of the XY model. The spin-spin correlation is analogous to the expectation value of Wilson's line integral[10] $\exp\left(i \sum_{\text{closed contour}} \phi(r,\hat{n})\right)$. The reader should recall that this line integral is the path integral expression equivalent to the computation of the static $q\bar{q}$ potential discussed earlier. In fact, choosing a closed rectangular contour

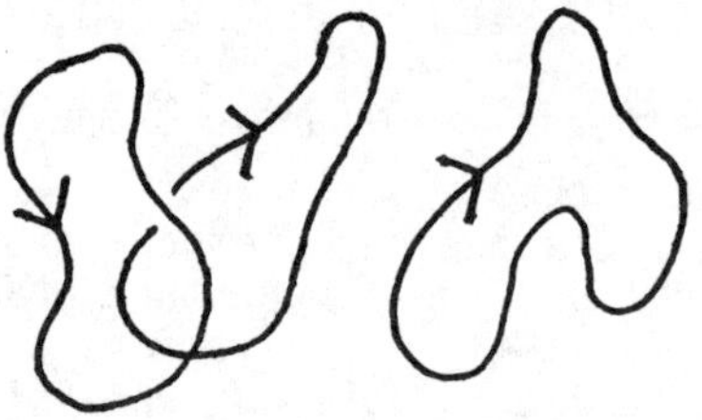

Fig. 5 Closed threads, $m_\mu(r)$

having spatial dimension R and temporal dimension T, then

$$V(R) = - \lim_{T\to\infty} \frac{1}{T} \ln \langle \exp[i \sum \phi(r,n)] \rangle \qquad (30)$$

Following the usual terminology of spin systems, if $\langle \exp[i \sum \phi(r,n)] \rangle$ is short-ranged in R (i.e. $V(R) \sim |R|$), then we say the gauge theory is "disordered", while if $\langle \exp[i \sum \phi(r,\hat{n})]$ is power-behaved (i.e. V(R) is short-ranged), then the theory is "ordered". The earlier discussion of the Abelian lattice gauge theory at finite temperature illustrates the general rationale behind these words.

Working from Eq. (29b) and inspired by the two dimensional XY model, a physical picture of the phases of Abelian lattice gauge theory results:

1. For $g < g_c$, the current loops $m_\mu(r)$ are small and irrelevant. The periodic character of the theory is irrelevant and it reduces to free electromagnetism. There is no quark confinement.
2. For $g > g_c$, the current loops are unbounded in size and are relevant. They disorder the system making the Wilson line integral correlation function fall at an exponential rate in R. Quark confinement follows.

This is an appealing physical picture. It also has many elements in common with some recent more general work by 't Hooft on quark confinement.[18] Unfortunately, the Partition function Eq. (29b) is difficult to use for quantitative calculations for several reasons. First, one must be able to enumerate closed current loops on the lattice. (It is possible to estimate the number of closed loops of a certain length from computer studies of constrained random walks.) Second, each element $m_\mu(r)$ of each loop interacts via a long range potential $V(r-r')$ with all the other elements $m_\mu(r')$. As in the lattice Coulomb gas problem, screening effects are essential in understanding the high temperature phase of this system.[17] Both of these problems have surfaced before in statistical mechanics--the counting problem appears in studies of dilute polymer solutions and the long range interactions between elements of closed threads appears in some models of the phases of ^{4}He.[19] One is led to consider constrained random walks in external fields which account for the long range interaction. The counting problem can then be written as a functional integral solution to the diffusion equation in an external field. M. Stone and P. Thomas[5] have carried out this program in an approximate fashion and have rewritten Eq. (29b) as scalar QED,

$$Z \sim \int d[A_\mu] d[\phi] d[\phi^*] \, e^{-\int [\frac{1}{4}F_{\mu\nu}^2 + (\nabla\phi)^2 + M^2\phi^2 + \lambda(\phi^*\phi)^2] d^4x} \qquad (31)$$

where M^2, the mass of the scalar field, is computed to be negative for large coupling g and positive for small g. This theory has two distinct phases: for $M^2 < 0$, a Higgs mechanism occurs and the vacuum supports the Meissner effect; for $M^2 > 0$, the vacuum is normal. If we were calculating the expectation value of the Wilson line integral in the Abelian lattice gauge theory, then after the duality trans-

transformations leading to the scalar QED formulation of the problem, we would be left with the calculation of the action of a pair of magnetic monopoles. But if the theory resides in the $M^2 < 0$ sector, the magnetic flux emanating from the poles forms flux tubes which generate a linear confinement potential. If $M^2 > 0$ there is no confinement. This is the correct picture of the phases of the theory, and the final reasoning is familiar, simple and reliable. It also bears out the intuitions of Mandelstam and 't Hooft who gave rough arguments mapping the problem of the confinement of colored quarks in QCD onto that of monopoles in a superconductor.[21] But Eq. (31) is not without difficulties. If the same approximations leading from Eq. (29b) to Eq. (31) are applied to the XY model in 3 dimensions one obtains scalar QED in 3 dimensions in place of Eq. (31). But this theory has a first order phase transition at its critical point while it is known from other analyses that the spin system has a second order transition. It would be useful to improve the arguments of ref. (5) and to pinpoint the source of the error.[21] Even though this approach is not quantitative in the immediate neighborhood of g_c, it is interesting and suggestive.

Using Eq. (31) we can consider lattice gauge theory at finite temperatures again. We must consider scalar QED in a heat bath. This problem has been studied in detail in other contexts and it was established that if the theory resided in the Higgs' phase (spontaneously broken symmetry) at low temperatures, then there will be a finite temperature T_c where the symmetry is restored.[22] In the language of the original lattice gauge theory this means that the theory will be quark confining up to a finite temperature T_c where the property will be abruptly lost. This is our earlier result now obtained from a different perspective.

CLOSING REMARKS

In all of the topics discussed in this review the concept of duality of statistical mechanics played an essential role. It is clear that this tool will be playing an increasingly important role in exact and approximate analyses of gauge theories. This powerful method, which predicted the critical point in the two dimensional Ising model years before Onsager solved it, will certainly be helpful in the quark confinement problem of QCD and, I believe, in our perception of gauge theories in general in the near future.

REFERENCES

1. A. M. Polyakov, "Thermal Properties of Gauge Fields and Quark Liberation", Trieste preprint, Oct. 1977.
2. L. Susskind, "Hot Quark Soup", SLAC preprint, Jan. 1978.
3. W. Fischler, J. B. Kogut and L. Susskind, "Quark Confinement in Unusual Environments", SLAC-PUB-2075, Jan. 1978.
4. T. Banks, R. Myerson and J. B. Kogut, Nucl. Phys. B129, 493 (1977).
5. P. R. Thomas and M. Stone, "Nature of the Phase Transition in a Nonlinear $O(2)_3$ Model", D.A.M.T.P. preprint 78/12.

6. P. Carruthers, reprinted in Collective Phenomena, Vol. I, p. 147 (1973); G. Baym and S. Chin, Phys. Lett. 62B, 241 (1976); G. Chapline and M. Nauenberg, Nature 264, 235 (1976).
7. J. C. Collins and M. J. Perry, Phys. Rev. Lett. 34, 1353 (1975). M. B. Kislinger and P. D. Morley, Phys. Letts. 67B, 371 (1977); ibid. 69B, 257 (1977); Phys. Rev. D13, 2765, 2771 (1976). B. A. Freedman and L. D. McLerran, Phys. Rev. D16, 1130, 1147, 1169 (1977).
8. J. B. Kogut and L. Susskind, Phys. Rev. D11, 395 (1975).
9. F. Wegner, J. Math. Phys. 12, 2259 (1971).
10. K. G. Wilson, Phys. Rev. D10, 2445 (1974); A. M. Polyakov, Phys. Lett. 59B, 82 (1975).
11. R. Savit, Phys. Rev. Lett. 39, 55 (1977).
12. L. Berezinskii, Zy. Eksp. Teor. Fiz. 59, 907 (1970) [JETP 32, 493 (1971)]; J. Villain, J. Phys. (Paris) 36, 581 (1975).
13. R. Myerson, Inst. for Adv. Study preprint, 1977.
14. R. Balian, J. M. Drouffe and C. Itzykson, Phys. Rev. D11, 2098 (1975).
15. H. E. Stanley and T. A. Kaplan, Phys. Rev. Lett. 17, 913 (1966).
16. J. M. Kosterlitz and D. J. Thouless, J. Phys. C6, 1181 (1973).
17. J. M. Kosterlitz, J. Phys. C7, 1046 (1974). J. V. José, L. P. Kadanoff, S. Kirkpatrick and D. R. Nelson, Phys. Rev. B16, 1217 (1977).
18. G. 't Hooft, "On the Phase Transition Towards Permanent Quark Confinement", Utrecht preprint, Dec. 1977.
19. See, for example, F. W. Wiegel, Phys. Reps. 16C, 57 (1975).
20. S. Mandelstam, in "Extended Systems in Field Theory", J. L. Gervais and A. Neveu, eds., Phys. Reps. 23C, 3 (1976). G. 't Hooft, in "High Energy Physics: Proceedings of the EPS International Conference, Palermo, June 1975", A. Zichichi, ed. (Editrice Compositori, Bologna, 1976).
21. Another analysis of Abelian lattice gauge theory has been made by M. Peskin, "Mandelstam-'t Hooft Duality in Abelian Lattice Models", HUTP-77/A083. He obtains results supporting Mandelstam-'t Hooft Duality and predicts a second order phase transition. His phase diagram, Fig. 2 of his preprint, suggests how approximate approaches to the problem can easily yield first rather than second order transitions. (I thank T. Banks for some correspondence on this issue.)
22. D. A. Kirzhnits and A. D. Linde, Phys. Lett. 42B, 471 (1972). S. Weinberg, Phys. Rev. D9, 3357 (1974). L. Dolan and R. Jackiw, Phys. Rev. D9, 3320 (1974).

Steps Toward the Heavy Quark Potential+

Frank Wilczek*

Princeton University
Princeton, New Jersey 08540

Abstract: Work on non-perturbative contributions to the heavy quark potential is reviewed. The position of this work in the program of establishing QCD, and the (major) steps needed to produce quantitative predictions in heavy quark spectroscopy, are assessed.

+Talk given at the Conference "50 Years of the Dirac Equation", Talahassee, Florida, April 1978.

*Supported in part by DOE Contract EY76-C02-3072 and by the A. P. Sloan Foundation.

ISSN: 0094-243X/78/030/$1.50

Asymptotically free gauge theories of the strong interactions[1-3] were first proposed in an attempt to understand Bjorken scaling[4] starting from horest field theory. As we shall hear from Ellis[5] and Politzer[5], these theories give quantitative predictions for deviations from scaling[6,7] which are beginning to have substantial experimental support[8]. Moreover, the demand of asymptotic freedom leads inevitably[9] to the color gauge theory of quarks and gluons (QCD)[1,3,10]. This theory has just those symmetries exhibited by the strong interactions[11,12] - C,P,T, the algebra of currents, flavor conservation - and SU(3) and chiral invariance with $(3\ \bar{3})$ breaking are readily incorporated. Other qualitative pieces of evidence are the success of quark models incorporating color gluon exchange interactions[13], the interpretation of Zweig's rule for heavy quarks as a "softening" of the strong interactions at short distances[14], and the whole success of the parton picture[15]. Finally there is the quantitative prediction for the rate of $\pi^0 \to \gamma\gamma$[16].

We therefore have a candidate fundamental theory for the strong interactions which is conceptually simple, very beautiful and symmetric, and which has had numerous qualitative and detailed quantitative[5,8] successes. Let us assume the predictions for scaling deviations continue to agree with more accurate and extensive future experiments. Can we then claim to have a satisfactory theory of the strong interaction?

The great obstacle to full acceptance of QCD is of course that the fundamental entities of the theory, quarks and gluons, are not the strongly interacting particles we see directly. Asymptotic freedom itself suggests that, since the coupling between quark decreases at short distances, it should increase at large distances, so that separating the quarks could become impossible. A much deeper hint of quark confinement is found in the strong-coupling limit of lattice gauge theories[17]. It may be very difficult to "prove" confinement directly - just as, for instance (for very different reasons), it is very difficult to show that matter crystallizes or becomes superconducting at low temperatures, starting from the Dirac equation and Coulomb forces. Similarly to what we have in these last cases, what we can hope for in the theory of strong interaction is a coherent picture of how the confined state might look and some quantitative results <u>starting</u> from this picture which can be checked against experiment.

Recently there have been very hopeful steps toward developing such a picture, in the work of Callan, Dashen, and Gross[18,19]. This work uses heavily tools forged by Polyakov[20] and 't Hooft[21]. A. Zee and myself[22] and then more systematically Callan, Dashen, Gross, Zee, and myself[23] have made some first tentative steps toward trying to extract quantitative results from the basic picture of confinement developed by CDG[18,19].

Although the mathematical apparatus used to develop the theory sometimes is formidable, the final picture can be understood in simple terms - with one major exception. The exception is that the picture has only been developed in Euclidean four-space, not Minkowski space-time. The Euclidean picture suffices to address the static quark potential, but for other problems, it is an important unsolved problem to understand the CDG picture in Minkowski space.

Feynman[24] has taught us that to evaluate quantities in field theory we need to do functional integrals over all field configurations, weighted by their action. Wilson[17] has taught us that the relevant quantity for static interactions between quarks in QCD is the loop integral

$$\lim_{T\to\infty} e^{-E(R)T} = \lim_{T\to\infty} \langle P e^{ig\oint A_\mu dx^\mu} \rangle \tag{1}$$

where E(R) is the energy between two static sources separated by a distance R, and on the right-hand side the loop integral is over a loop of width R and length T. The gauge potential A_μ is a matrix so ordering along the loop is required. The average on the right-hand side is to be taken over all field configurations weighted by their action.

The CDG picture is a (well-motivated) guess about which configurations are the most important ones. In perturbation theory in the zeroth order we have Coulomb fields between the quarks, and higher-order corrections are due to exchange of transverse gluons - plane-wave configurations, typically of high frequency. For small coupling, we may say roughly that in perturbation theory we have typically small-amplitude, high frequency fluctuations of the fields A_μ giving the corrections. In QCD there is another kind of configuration which CDG argue is more important (except for very small R). These are the instanton configurations[25]. These configurations are large-amplitude ($\sim\frac{1}{g}$) but typically smooth, low-frequency, coherent fluctuations of the fields A_μ. It has been argued[19] that inside hadrons these fluctuations are rare enough that it makes sense to treat them as non-overlapping, a sort of gas of pseudoparticles. Then an important ingredient in calculating the interaction energy between quarks becomes to calculate the force induced by such fluctuations.

For phenomenological purposes we need not only the force between static sources but also the dependence of the interaction on spins, etc. A. Zee and I[22] proposed a generalization of Equ. (1) including spins, and then CDGWZ[23] and C. de Carvalho[26] evaluated the full potential, using the Foldy-Wothuysen expansion, to order $\frac{1}{m^2}$, where m is the quark mass. The result is an explicit but very complicated-looking expression, which you can find in Ref. (23).

This calculation is clearly an incomplete calculation, for the following reasons:

i) It gives the result for a single instanton, and must be folded into the space-time distribution of instantons. According to Ref. (1 this distribution is highly non-uniform and non-trivial to calculate.

ii) We have assumed fixed sources. Inside hadrons, quarks certainly move, and for instance spatially large, slow fluctuations will be cut off because they see the average charge of the quark-antiquark pair, which is zero. This effect is not likely to be large for heavy quarks, but it is probably very important for light quarks. This is where a Minkowski-space picture would be very helpful.

iii) Even in purely heavy quark systems, we need to take into accoun the dynamical effects of virtual light quarks, and of mixing and annihilation.

Evidently there is a great deal to be done. However, it seems that each of the above problems is a finite, well-defined problem and one can expect each to be solved in turn. In the end, assuming all this is done, will we have improved very much on simple quark or bag models? From a phenomenological point of view, probably not. These models already give 10 - 20 % accuracy for most static properties of hadrons, and it is hard to imagine doing much better any time soon, since higher-order terms of order a few percent are dropped at several points. However, since we have many experimental quantities to fit (spin-splittings, radiative and pionic transitions, radial splittings, ...) very few parameters, and a picture very different in detail from previous ones (rapid spatial dependence of interactions on quark positions, weaker couplings, ...) even this good a fit would be impressive. One can dream of understanding the odd properties of the pseudoscalar mesons η, η', η_c or especially of finding some unexpected new effect. But the crucial result will be (if it works) that a physically consistent picture of confinement has emerged. I think this would remove the last serious barrier to full acceptance of QCD, otherwise so attractive, as the theory of strong interactions.

References

1) D. Gross, F. Wilczek, Phys. Rev. Lett. 30, 1343 (1973); Phys. Rev. D8, 3633 (1973).

2) H. D. Politzer, Phys. Rev. Lett. 30, 1346 (1973).

3) S. Weinberg, Phys. Rev. Lett. 31, 494 (1973).

4) J. Bjorken, Phys. Rev. 179, 1547 (1969).

5) These proceedings.

6) D. Gross, F. Wilczek, Phys. Rev. D9, 980 (1974).

7) H. Georgi, H. Politzer, Phys. Rev. D9, 416 (1974).

8) Bosetti et al., "Analysis of Nucleon Structure Functions in CERN Bubble Chamber Neutrino Experiments," Oxford Preprint 16/78, to be published in Nucl. Phys. B.

9) S. Coleman, D. Gross, Phys. Rev. Lett. 31, 851 (1973).

10) H. Fritzsch, M. Gell-Mann, H. Leutwyler, Phys. Lett. 47B, 365 (1973).

11) The numerous contributions of M. Gell-Mann to elucidating these symmetries must be explicitly remarked.

12) Actually there is a problem with "naturally" incorporating the P and T invariance:

R. Peccei, H. Quinen, Phys. Rev. D16, 1791 (1977)

F. Wilczek, Phys. Rev. Lett. 40, 279 (1978)

S. Weinberg, Phys. Rev. Lett. 40, 223 (1978)

However, like chiral and SU(3) invariance, this probably should not be considered a problem in QCD, but in weak (or even gravitational?) interactions.

13) A. Chodos et al., Phys. Rev. D10, 2599 (1974); Phys. Rev. D12, 2060 (1975); and others.

14) T. Appelquist, H. Politzer, Phys. Rev. 34, 43 (1975).

15) R. Feynman, "Photon-Hadron Interactions", (Benjamin, 1973).

16) S. Adler, Phys. Rev. 177, 2426 (1969); J. Bell, R. Jackiw, Nuovo Cimento 60A, 47 (1969).

17) K. Wilson , Phys. Rev. D10, 2445 (1975).

18) C. Callan, R. Dashen, D. Gross, Phys. Rev.

19) C. Callan, R. Dashen, D. Gross "A Theory of Hadronic Structure", IAS Preprint (June 1978).

20) A. Polyakov, Nucl. Phys. B120, 429 (1977).

21) G. 't Hooft, Phys. Rev. D14, 3432 (1976).

22) F. Wilczek and A. Zee, Phys. Rev. Lett. 40 83 (1978).

23) C. Callan, R. Dashen, D. Gross, F. Wilczek, A. Zee, "Effect of Instantons on the Heavy Quark Potential", IAS Preprint (1978).

24) R. Feynman, Rev. Mod. Phys. 20, 267 (1948).

25) A. Belavin, A. Polyakov, A. Schwartz, Y. Tyupkin, Phys. Lett. 59B, 85 (1975).

26) C. de Carvalho, to be published.

NEUTRAL CURRENTS AND GAUGE THEORIES --PAST, PRESENT, AND FUTURE

J. J. Sakurai
Department of Physics, University of California,
Los Angeles, California 90024

Presented at "Current Trends in the Theory of Fields--50 Years of the Dirac Equation--In Honor of Paul A. M. Dirac" (April 6-7, 1978, Florida State University, Tallahassee, Florida).
To be published in AIP Proceedings.

TABLE OF CONTENTS

ISSN: 0094-243X/78/038/$1.50

NEUTRAL CURRENTS AND GAUGE THEORIES --PAST, PRESENT, AND FUTURE*

J. J. Sakurai
University of California, Los Angeles, CA 90024

It is a pleasure to have this opportunity to address a distinguished audience gathered here to celebrate the 50th anniversary of the relativistic wave equation. Anybody who has followed the development of weak interaction physics in the past half century must by now realize that the power of relativistic quantum theory, pioneered by Professor Dirac,[1] in making nontrivial predictions is by no means limited to the narrow domain of QED. What is even more striking, theorists nowadays speculate on the unification of weak and electromagnetic forces while experimentalists, within the next decade or two, will explore the energy regions where weak and electromagnetic effects are bound to be comparable and, in a certain sense, inseparable. In my talk I would like to review our past accomplishments, report on the present status and, at the same time, indicate possible future directions in this fascinating field of neutral currents and gauge theories.

I. PREHISTORY

As is well known, the neutrino was conceived by Pauli[2] and delivered by Fermi in 1933 or 1934, depending on whether you read Italian[3] or German.[4] Fermi's beta decay interaction

$$n \to p + e^- + \bar{\nu} \tag{1.1}$$

is the first example of what later came to be known as the charged-current interactions [see Fig. 1(a)]. It is also appropriate to

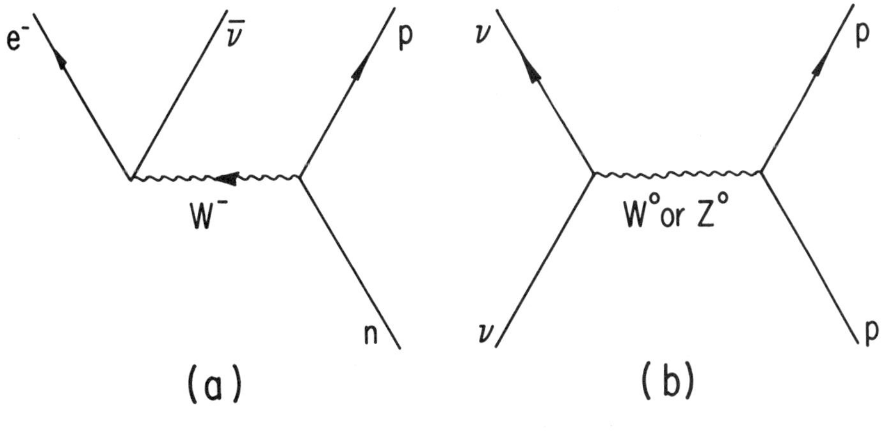

Fig. 1. Examples of the charged-current (a) and the neutral-current (b) interactions.

*Supported in part by the National Science Foundation.

mention, on this festive occasion, that we find in Fermi's theory the earliest application, outside QED, of second quantization proposed by Professor Dirac[5] and others.[6]

Yukawa usually gets credits for inventing mesons to account for nuclear forces. Actually, in his 1935 paper,[7] he was a bit too ambitious; he intended his "U quantum" to play the dual role of being the carrier of both strong forces (nucleon-nucleon interactions) and weak forces (beta decay interactions). In other words his U quanta corresponded, in modern notation, to weak intermediate bosons ($W^{\mp}$) as well as to mesons ($\pi^{\pm}$, $\rho^{\pm}$, ...).

Both Yukawa and Fermi were motivated by analogy with the electromagnetic interactions. Fermi emphasized in his papers that the emission of an $e^{-}\bar{\nu}$ pair is like photon emission; in his own words,

> "...in modo analogo alla formazione di un quanto di luce."

In fact, even though he was completely aware of the possibility of writing down the most general four-fermion interaction of the form $(\bar{e}\Gamma\nu)(\bar{p}\Gamma n)$ where Γ may stand for any of the 16 Dirac matrices, guided by analogy with electromagnetism, he argued that it is natural to let $\Gamma \to \gamma_\lambda$ so that we end up with a vector-type interaction. It is amusing that Fermi's 1933-1934 interaction is essentially correct even today; all we need to do is insert[8] $1 + \gamma_5$ and replace the nucleon fields by the Cabibbo-rotated[9] (or Gell-Mann-Lévy-rotated[10]) quark fields.

Like Fermi, Yukawa was also motivated by analogy with electromagnetism. He starts the main part of his 1935 paper by saying[7]

> "In analogy with the scalar potential of the electromagnetic field a function U(x,y,z,t) is introduced to describe the field between the neutron and the proton."

After the publication of the papers of Fermi and Yukawa, it might have appeared almost irresistible to make this analogy with electromagnetism even deeper by arguing: (i) It is natural to consider "neutral current" interactions mediated by a neutral boson W° (or Z°) [see Fig. 1(b) for an example of the neutral-current interactions]; after all the photon is neutral. (ii) It is natural to construct a gauge theory of weak interactions; after all, QED is based on the gauge principle. So you see how this discussion of "prehistory" motivates the title of my talk, "Neutral Currents and Gauge Theories."

In spite of all this, besides World War II that slowed down the progress of pure science, there were two stumbling blocks which had to be overcome before theorists started thinking seriously about neutral currents and/or gauge theories. First of all, it was impossible to study neutral currents in nuclear decay transitions--$\nu\bar{\nu}$ emission processes would be completely overshadowed by radiative processes due to the much stronger electromagnetic interactions; to study neutral currents we had to have reactions initiated by good neutrino "beams," which did not become available until the middle '50s for reactor (anti-) neutrinos[11] and until the early '60s for accelerator neutrinos.[12] Second, until about 1957 physicists--theorists and experimentalists alike--were misled by wrong experiments and believed that the beta-decay interaction involved STP, a combination that does not permit the construction of weak-interaction gauge theories in the usual sense.

II. HISTORY

In my view the history of neutral currents and gauge theories started around 1957, but this may have to do with the fact that I started doing research around that time.

The first serious attempt to unify the weak and electromagnetic interactions was made in the 1957 Annals of Physics paper of Schwinger.[13] In that paper he introduced a charge triplet of spin-one fields; according to him the charged members of the triplet are responsible for the charged-current weak interactions while the neutral member is to be identified with the photon. All this was proposed before the V-A structure of the charged-current interactions was established, and with his characteristic ingenuity he somehow managed to obtain a tensor-type (rather than axial-vector-type) interaction for the Gamow-Teller part of nuclear beta decay.

In 1958 Bludman[14] constructed a weak-interaction model that includes neutral currents, based on the Yang-Mills gauge principle.[15] Like Schwinger he had a triplet of spin-one fields, $W^{\pm,0}$; however, unlike Schwinger, the neutral member of Bludman's triplet had nothing to do with the Maxwell field; it was, in fact, coupled to weak neutral currents. By that time the V-A structure[8] of the charged-current interactions had been established, and it was natural for Bludman to generate the entire interaction via the gauge derivative substitution à la Yang and Mills

$$\partial_\mu \to \partial_\mu - ig\, \underset{\sim}{T}\cdot\underset{\sim}{W}_\mu$$

for left-handed fermions where $\underset{\sim}{T}$ is the weak-isospin matrix whose 1±i2 component is known from the charged-current interactions. In this manner he had an SU(2) symmetric gauge theory in which the weak boson masses that break local gauge symmetry were put in "by hand." Note that in his scheme both the neutral-current and charged-current interactions are of the pure V-A type.

The group $SU(2) \otimes U(1)$, as used today in unified models of weak and electromagnetic interactions, was first introduced by Glashow whose attempt was reported briefly in Gell-Mann's 1960 Rochester Conference talk.[16] Suppose we try to write down the lepton analog of the Gell-Mann-Nishijima formula

$$Q = I_3 + (S+B)/2 \tag{2.1}$$

We may consider

$$Q = T_3 + Y \tag{2.2}$$

where $\underset{\sim}{T}$ stands for weak isospin such that $T_\pm$ appears in the charged-current V-A interactions, and Y, "weak hypercharge" (not to be confused with the strong hypercharge S+B). In the neutrino-electron sector the matrix representations of $\underset{\sim}{T}$ and Y can be written as follows:

$$\begin{array}{c} \quad\; \nu_L \;\;\; e_L^- \;\;\; e_R^- \\ \underset{\sim}{T} = \left(\begin{array}{cc|c} \multicolumn{2}{c|}{\frac{1}{2}\underset{\sim}{\tau}} & \begin{array}{c}0\\0\end{array} \\ \hline 0 & 0 & 0 \end{array}\right) \\ \\ Y = \begin{pmatrix} -\frac{1}{2} & 0 & 0 \\ 0 & -\frac{1}{2} & 0 \\ 0 & 0 & -1 \end{pmatrix} \end{array} \tag{2.3}$$

where $\underset{\sim}{\tau}$ is the usual 2 × 2 Pauli matrix. Note

$$[T_\alpha, T_\beta] = i\varepsilon_{\alpha\beta\gamma} T_\gamma \quad ,$$
$$[T_\alpha, Y] = 0 \; , \tag{2.4}$$

hence $SU(2) \otimes U(1)$.

The 1961 Nuclear Physics paper of Glashow,[17] entitled "Partial-symmetries of Weak Interactions," already contains much of what later came to be known as the "Weinberg-Salam model." Like Bludman he had a $\underset{\sim}{W}_\lambda$ triplet ($\underset{\sim}{Z}_\lambda$ triplet in Glashow's notation) coupled to the weak-isospin current J_λ^α (α = 1,2,3); in addition he introduced a singlet spin-one field B_λ (Z_λ^s in Glashow's notation) coupled to the weak-hypercharge current J_λ^Y. Glashow breaks local gauge symmetry by inserting, in the basic Lagrangian, off-diagonal as well as diagonal mass terms for these gauge fields by hand, and, as the mass matrix gets diagonalized, one of the linear combinations of the W_λ^3 and B_λ becomes the massless photon field A_λ coupled to

$$J_\lambda^{em} = J_\lambda^3 + J_\lambda^Y \tag{2.5}$$

while the other orthogonal linear combination Z_λ (B_λ in Glashow's notation) remains massive; specifically we have

$$\begin{aligned} A_\lambda &= B_\lambda \cos\theta + W_\lambda^3 \sin\theta \quad , \\ Z_\lambda &= W_\lambda^3 \cos\theta - B_\lambda \sin\theta \quad , \end{aligned} \tag{2.6}$$

as

$$\begin{array}{cc} B^0 & W^0 \\ \left(\begin{array}{c|c} m_B^2 & -m_{BW}^2 \\ \hline -m_{BW}^2 & m_W^2 \end{array}\right) \end{array} \longrightarrow \begin{array}{cc} \gamma & Z \\ \left(\begin{array}{c|c} 0 & 0 \\ \hline 0 & m_Z^2 \end{array}\right) \end{array} \; . \tag{2.7}$$

In this model the source of Z_λ is proportional to a linear combination of $J_\lambda{}^3$ and J_λ^{em},

$$J_\lambda{}^3 - \sin^2\theta \; J_\lambda^{em} \tag{2.8}$$

where the angle θ is what is now known as the Weinberg angle θ_W. (Strictly speaking the angle θ in Glashow's original paper is $\pi/2 - \theta$ here.)

In 1964 Salam and Ward,[18] being apparently unaware of Glashow's paper, rediscovered Glashow's 1961 model and correctly concluded that the appearance of the neutral current is "about the minimum price we must pay to achieve the synthesis of weak and electromagnetic interactions." They were, however, discouraged by the wrong CERN experiment to be mentioned later.

There were other papers on W^0 bosons and neutral currents motivated by attempts to obtain the $\Delta I = 1/2$ rule in nonleptonic decays.[19] I'll not review them here because they are not directly related to the line of development I am following.

III. EARLY EXPERIMENTS

Already in the early '60s the importance of looking for neutral currents experimentally was clearly recognized by some theorists. In 1960, as the AGS at Brookhaven and the PS at CERN were about to come in operation, Lee and Yang[20] listed nine reasons for doing accelerator neutrino experiments. The "possible existence of a neutral lepton current" was included in the list. They emphasized, in particular, the importance of studying

$$\nu + p \rightarrow \nu + p \; , \tag{3.1}$$

a reaction for which we had no positive evidence until 16 years later.

In 1961, when electron-positron colliding-beam facilities were being planned at Orsay and Frascati, Cabibbo and Gatto[21] discussed possible experiments that could be performed at such facilities. One of the various experiments mentioned was a study of weak-electromagnetic interference in

$$e^+ + e^- \rightarrow \mu^+ + \mu^- \tag{3.2}$$

expected if the weak interactions contained a neutral-current piece. The interference effect discussed by Cabibbo and Gatto will be detected for the first time probably in 1979 at PETRA. So you see that the time scale of weak interactions is impressively long.

Even in the '60s there were experiments that came dangerously close to discovering neutral currents. Possible examples of neutral current events could be found in the very first accelerator neutrino experiment carried out in 1962 by a Columbia-Brookhaven group[12]--Lederman, Schwartz, Steinberger, and collaborators. As is well known, they unambiguously observed about fifty events with long muon tracks due to the ordinary charged-current interactions, e.g.,

$$\nu_\mu + n \rightarrow \mu^- + p \ , \qquad \mu^- + p + \pi^o \ . \tag{3.3}$$

However, there were also shower and prong events with no obvious muon tracks. By studying the energy distribution of the electron showers they convinced themselves that the shower events could not possibly be due to reactions like

$$\nu_e + n \rightarrow e^- + p \ , \qquad \nu_e + p \rightarrow e^- + p + \pi^+ \ , \tag{3.4}$$

after all the primary purpose of the experiment was to test $\nu_\mu \neq \nu_e$ --but they did not examine critically whether these shower and prong events were caused by an uninteresting neutron contamination in the beam or they represented something more interesting, viz. neutrino-induced neutral-current reactions:

$$\begin{aligned} &\nu + p \rightarrow \nu + p \ , \qquad \nu + p + \pi^o \\ &\nu + n \rightarrow \nu + p + \pi^- \ . \end{aligned} \tag{3.5}$$

Accelerator neutrino physics was studied also in Europe, in particular, by the CERN 1.2 m bubble chamber group in 1963-1964. There was a graduate student of Perkins at Oxford, E. C. M. Young, whose job was to examine the neutron background in that experiment. In his thesis he is reputed to have concluded that the number of muonless events was about three times the neutron background he could estimate. In fact, it was mentioned in an unpublished memorandum of Perkins that if these excess muonless events had been attributed to neutral currents, we would have obtained a neutral-to-charged current ratio of 17 ± 6%, roughly the currently accepted value. Yet at the 1963 Siena Conference the group reported for νp elastic scattering

$$\sigma(\nu p \rightarrow \nu p)/\sigma(\nu n \rightarrow \mu^- p) < 3\% \tag{3.6}$$

(to be compared with 10-20% indicated by present-day data), a very curious result not withdrawn until 1970[22] when theorists all over the world really started screaming for neutral currents. This low limit discouraged many theorists, including Salam and Ward,[18] from considering seriously models where neutral currents and charged currents participate in the hadron sector with about the same strength. I was told that there were even theoretical papers rejected by The Physical Review on account of their incompatibility with the CERN limit (3.6).

No discussion of near misses is complete without mentioning the 1970-1971 SLAC beam dump experiment--commonly known as the "Black Hole

Experiment"--carried out by Schwartz, Wojcicki, Dorfan, and collaborators. In this experiment, described in A. F. Rothenberg's Stanford thesis,[23] particles produced in the SLAC beam dump area were allowed to go through nearly 200 ft of dirt and rocks. A spark chamber complex was set up to look for the interactions of possible neutral particles that survived the ordeal. There were unmistakable examples of neutrino-induced events with long muon tracks. However, there were also events with no muons. Two tantalizing examples of this latter type are shown in Fig. 2, kindly supplied to me by Dave Dorfan of UC Santa Cruz. In hindsight these events may well have been due to the neutral-current interactions. One may now argue that the probability of high-energy neutrons entering the chamber horizontally in an arrangement of this kind is essentially negligible, but at that time no serious attempts were made to interpret them as neutral-

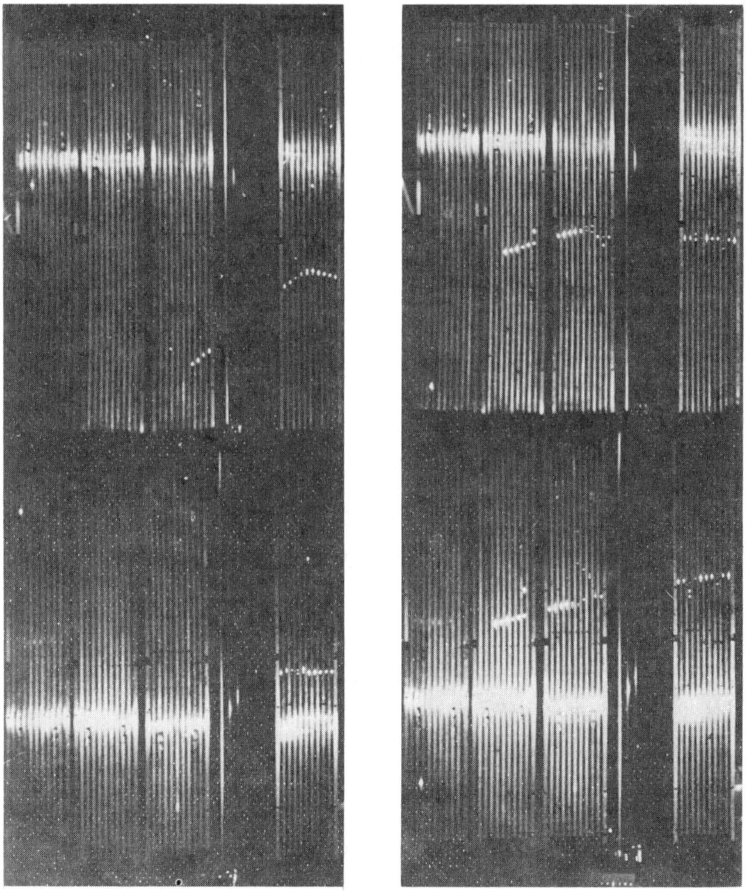

Fig. 2. Two possible examples of pre-1973 neutral-current events in the SLAC beam dump experiment. The beam direction is from the left to the right.

current events. In fact these pictures are published here for the first time.

Why am I mentioning these early experiments? What do we learn from studying the past? Answer: If we don't have the right theoretical prejudice, progress in experimental physics is likely to be very slow.

IV. DAWNING OF THE MODERN ERA--THEORETICAL

It is fair to say that for most theorists the modern era really started with Weinberg and Salam: the 1967 paper of Weinberg published in Physical Review Letters[24] and Salam's report[25] at the Sixth Nobel symposium held at Aspenäsgården, Lerum, in May, 1978. With Weinberg and Salam a new element was added to the 1961 Glashow model: mass generation via spontaneous breakdown of gauge symmetry. It is not my purpose here to discuss in detail how the so-called "Higgs mechanism" actually operates but I would like to call your attention to Fig. 3, taken from the September 1977 issue of CERN Courier.

The Higgs bosons are the origin of 'spontaneous symmetry breaking'. Abdus Salam has a personal picture to communicate this abstruse concept. Imagine a banquet where guests sit at round tables. A bird's eye view of the scene presents total symmetry, with serviettes alternating with people around each table. A person could equally well take a serviette from his right or from his left. The symmetry is spontaneously broken when one guest decides to pick up from his left and every one else follows suit.

Fig. 3. Salam's analogy.

I should also mention that many workers, besides Higgs,[26] contributed to what is now known as the Higgs mechanism; a partial list may include Englert, Brout, and Kibble.[27]

As far as the practical consequences are concerned, everything follows once we introduce a Higgs scalar doublet ($T = 1/2$, $Y = 1/2$)

$$\phi = \begin{pmatrix} \phi^+ \\ \phi^o \end{pmatrix} \tag{4.1}$$

such that, when the gauge symmetry is broken spontaneously, ϕ^0 acquires a nonvanishing vacuum expectation value. Since the scalar doublet introduced has weak isospin as well as weak hypercharge, there are couplings with the gauge bosons of the form $\phi^\dagger\phi(V_\lambda)^2$--analogous to the $e^2\phi^\dagger\phi(A_\lambda)^2$ term that appears in the electrodynamics of charged scalar fields--where V_λ may stand for $W_\lambda^{\pm,0}$ or $B_\lambda{}^0$. A nonvanishing vacuum expectation value for ϕ^0 then implies that the gauge fields, $W_\lambda^{\pm,0}$ and $B_\lambda{}^0$, now acquire finite masses.[28] The structure of the mass matrix for the B^0W^0 complex is such that one of the eigenvalues is guaranteed to vanish; the reason is that $m_W{}^2$, $m_B{}^2$, and m_{BW}^2 in (2.7) are now proportional to g^2, g'^2, and gg', respectively, where g and g' are the coupling constants of $\underset{\sim}{W}_\lambda$ and B_λ. The rest of the theory is the same as the 1961 Glashow model.

When the symmetry is broken in this particular manner, there are definite relations for the weak boson masses, already obtained in Weinberg's original paper[24]:

$$m_W = 37.4 \text{ GeV}/\sin\theta_W \;, \qquad m_Z = m_W/\cos\theta_W \;. \tag{4.2}$$

These relations can be easily derived using

$$\begin{aligned} m_W{}^2 &= \frac{1}{4}\, g^2 |\langle\phi\rangle_{vac}|^2 \;, \\ G/\sqrt{2} &= g^2/8\; m_W{}^2 \;, \qquad e/g = \sin\theta_W \;\; \text{etc.} \end{aligned} \tag{4.3}$$

Nowadays the proof of (4.2) is taught to students taking QFD 101. Note that θ_W that appears in (4.2) can be measured by performing low-energy neutral-current experiments because Z is coupled to the current (2.8). I'll come back to the experimental implications of these remarkable mass relations later on.

It is worth mentioning that the spontaneously broken gauge theory (hereafter denoted by SBGT) has an added bonus: good high-energy behavior. As a result, the theory can be shown to be renormalizable; we have definite prescriptions for calculating higher-order weak amplitudes. Here again many people contributed to the subject but the major credit should go to G. 't Hooft, who, as a 24-year-old graduate student under Veltman, wrote in 1971 two fundamental papers published in Nuclear Physics.[29,30]

For those who have no strength to face the ghosts of Popov and Fadeev needed for the renormalization program, let me mention a low-brow argument in support of SBGT based on "tree unitarity," due to Weinberg, Bell, Llewellyn Smith, and Cornwall et al.[31]

Consider

$$\nu + \bar{\nu} \to W^- + W^+ \ . \tag{4.4}$$

This reaction is allowed by the charged-current interactions via electron exchange [see Fig. 4(a)]. However, if this were the only interaction, the cross section would go like[32]

$$\sigma \simeq G^2 s/12\pi \tag{4.5}$$

at high energies, eventually violating unitarity. This bad high-energy behavior would be cancelled with an additional diagram due to the propagation of the Z boson in the s channel [see Fig. 4(b)] provided the couplings of Z to $\nu\bar{\nu}$ and W^+W^- are precisely of the form dictated by SBGT. Furthermore, if this kind of argument is repeated for

$$e^- + e^+ \to W^- + W^+ \ , \tag{4.6}$$

it is possible to show that the best high-energy behavior is attained when, in addition to Z and γ, we also include the propagation of the Higgs boson in the s channel again with the coupling strength that follows from SBGT.

Despite all these attractive features the model was actually facing in the late '60s a seemingly insurmountable hurdle. This had

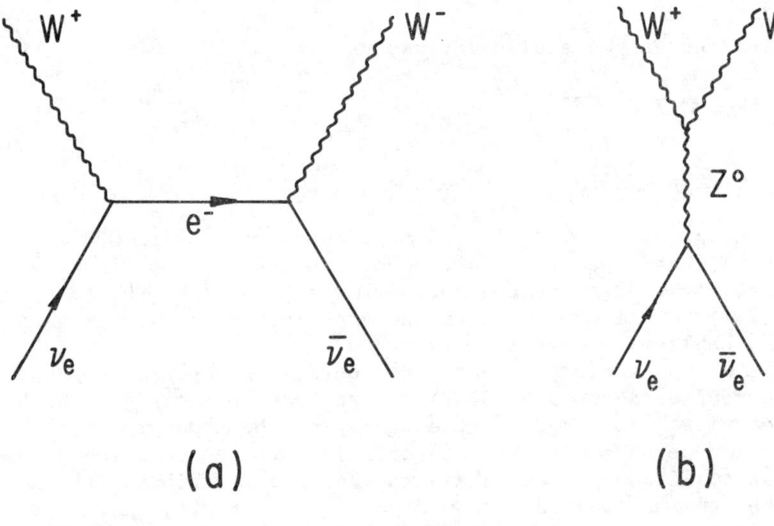

Fig. 4. Diagrams for $\nu + \bar{\nu} \to W^- + W^+$

to do with the problem of strangeness-changing neutral currents. Consider any model (not necessarily SBGT à la Salam and Weinberg) where the neutral-current source Q_3 is generated by commuting the charged-current sources $Q_\pm$:

$$[Q_+, Q_-] = 2Q_3 \ . \tag{4.7}$$

Suppose we substitute in (4.7) what was believed in the late '60s to be the correct charged-current source of the hadronic world

$$Q_+ = (Q_-)^\dagger = \frac{1}{2} \int d^3x \, [u^\dagger(1+\gamma_5)d \cos\theta_c + u^\dagger(1+\gamma_5)s \sin\theta_c] \tag{4.8}$$

where θ_c stands for the Cabibbo angle. We then obtain a strangeness-changing piece in the neutral current source

$$\cos\theta_c \sin\theta_c \int d^3x \, [s^\dagger(1+\gamma_5)d + d^\dagger(1+\gamma_5)s] \ . \tag{4.9}$$

Yet, experimentally, the strangeness-changing neutral current appears to be absent or very highly suppressed[33]:

$$\begin{aligned} &\Gamma(K_L \to \mu^+\mu^-)/\Gamma(K^+ \to \mu^+\nu_\mu) \simeq (4\pm1) \times 10^{-9} \qquad [O(\alpha^4)] \\ &\Gamma(K^+ \to \pi^+\nu\bar{\nu})/\Gamma(K^+ \to \pi^o e^+\nu) < 1.2 \times 10^{-5} \ . \end{aligned} \tag{4.10}$$

For this reason, until the early '70s, the gauge theory of weak interactions looked to many people like an interesting curiosity which perhaps may be applicable to the leptonic world--it is no coincidence that the title of the 1967 paper of Weinberg was "A Model of Leptons"--but not a realistic proposal that describes the weak interactions of hadrons as well.

The situation changed dramatically with the 1970 suggestion of Glashow, Iliopoulos, and Maiani.[34] They proposed that a fourth quark[35] c enter in the weak interactions so that the hadronic weak currents are to be built up of two left-handed doublets:

$$\begin{pmatrix} u \\ d\cos\theta_c + s\sin\theta_c \end{pmatrix}, \qquad \begin{pmatrix} c \\ -d\sin\theta_c + s\cos\theta_c \end{pmatrix}. \tag{4.11}$$

When $Q_\pm$ are formed with the two doublets (4.11), we readily see that the resulting Q_3 no longer contains the undesirable strangeness-changing piece (4.9).

I have no time here to recount the success story of the GIM scheme starting with the November (1974) Revolution[36]--ψ/J (= $c\bar{c}$), $D^{+,0}$ (= $c\bar{d}$, $c\bar{u}$), F^+ (= $c\bar{s}$) and all that--but I can only concur with John Ellis[37] who reviewed this subject at Cargèse last summer with the following remark:

> "We would conclude that the GIM scheme is so well established that it is almost ripe to be put in the text book."

V. DAWNING OF THE MODERN ERA--EXPERIMENTAL

Just as the modern era for theorists started with the papers of Weinberg and Salam in 1967-1968 the modern era for experimentalists started in the summer of 1973 with the discovery of neutrino-induced muonless events, most readily interpretable as

$$\overset{(-)}{\nu}_{\mu} + N \rightarrow \overset{(-)}{\nu}_{\mu} + \text{hadrons} \; . \tag{5.1}$$

The great discovery was made by the Gargamelle Collaboration[38] made up of seven bubble-chamber groups in Western Europe. In this experiment a freon-filled chamber was exposed to focused high-energy neutrino and antineutrino beams from the PS at CERN. The peak energies of these beams were in the 1-2 GeV range.

The major advantage of this neutrino experimental over the earlier 1.2 m chamber experiment was the sheer dimension of the chamber. To see how appropriately this French-funded chamber was named, we consult Pierre Larousse's Grand dictionnaire universel du XIX$^{\underline{e}}$ siècle:

> "GARGAMELLE, femme de Grandgousier et mère de Gargantua, dans le livre de Rabelais; femme aux proportions colossales et telle qu'on peut se la figurer pour avoir donné le jour à celui qui est resté le type, dans le langage vulgaire, du mangeur dont rien ne peut rassasier l'appétit gigantesque."

More seriously, the chamber, filled with 18 tons of freon (CF_3Br), is 4.8 m long, substantially longer than the inelastic mean free path for hadrons (0.7 m) and the radiation length (0.1 m). Because of this the Gargamelle Collaboration was able to study the spatial distribution of neutral-current events in the direction parallel to the incident neutrino beam over a distance considerably larger than the hadron mean free path. If the observed muonless events had been neutron-induced, we would have expected the ratio of the muonless events to the charged-current events to attenuate strongly as a function of distance. The spatial distribution of the observed muonless events were, in fact, uniform just as that of the charged-current events.

The ratios of the neutral to charged current events, denoted by $R_{\nu N}$ and $R_{\bar{\nu}N}$, were reported to be

$$\begin{aligned} R_{\nu N} &\equiv \sigma(\nu + N \rightarrow \nu + \text{hadrons})/\sigma(\nu + N \rightarrow \mu^- + \text{hadrons}) = (23\pm3)\% \; , \\ R_{\bar{\nu}N} &\equiv \sigma(\bar{\nu} + N \rightarrow \bar{\nu} + \text{hadrons})/\sigma(\bar{\nu} + N \rightarrow \mu^+ + \text{hadrons}) = (46\pm9)\% \; . \end{aligned} \tag{5.2}$$

Note that the strength of the neutral-current couplings is quite comparable to that of the charged-current couplings.

This discovery was quickly confirmed at NAL by a Harvard-Pennsylvania-Wisconsin collaboration who studied the interactions of neutrinos (a "natural" mixture of ν and $\bar{\nu}$) of considerable higher energies in a calorimeter-spark-chamber complex. However, because of an "extremely complex set of historical developments" (in David Cline's own words) this experiment, already mentioned in Myatt's review talk[39] at the Bonn Conference held in August, 1973, did not appear in print in Physical Review Letters[40] until April, 1974.

By the London Conference held in July, 1974, the inclusive neutral-current reactions (5.1) had also been detected by a Caltech group.[41] With or without other supportive evidence from single-pion production[42] and $\bar{\nu}_\mu e$ scattering,[43] there was no shadow of doubt in anybody's mind that the neutral currents were here to stay. To me the most convincing thing about the three inclusive experiments was the following. The three groups used very different detection devices and very different neutrino beams; the background problems they had to face were very dissimilar. Furthermore they have very different mental attitudes towards experimental physics. Yet the final numbers obtained by the three groups more or less agreed with each other within errors.

In addition to the inclusive reactions (5.1), between 1973 and now, the neutral-current interactions were established in the following exclusive reactions

$$\overset{(-)}{\nu} + p \to \overset{(-)}{\nu} + p \ , \tag{5.3a}$$

$$\overset{(-)}{\nu} + p \to \overset{(-)}{\nu} + p + \pi^o \ , \quad \overset{(-)}{\nu} + n + \pi^+ \ , \tag{5.3b}$$

$$\overset{(-)}{\nu} + n \to \overset{(-)}{\nu} + n + \pi^o \ , \quad \overset{(-)}{\nu} + p + \pi^- \ , \tag{5.3c}$$

$$\overset{(-)}{\nu}_\mu + e^- \to \overset{(-)}{\nu}_\mu + e^- \ , \tag{5.3d}$$

$$\bar{\nu}_e + e^- \to \bar{\nu}_e + e^- \ . \tag{5.3e}$$

Attempts have also been made to look for parity violation due to the neutral-current interactions in radiative transitions between atomic levels. I'll come back to these topics later.

VI. DETERMINATION OF NEUTRAL-CURRENT COUPLINGS--THE NEUTRINO-HADRON SECTOR

Since the 1973 Gargamelle discovery there have been essentially two approaches to neutral-current phenomenology. In the first approach, followed by most experimental rapporteurs in major international conferences, one analyzes available data--$R_{\nu N}$, $R_{\bar{\nu}N}$, etc.--within the framework of some specific theoretical model--the one-parameter Salam-Weinberg model, in particular. In the second approach, on the other hand, one attempts to look at the data using the most general phenomenological framework compatible with invariance principles; the basic questions asked are: What are the observables in the neutral-current interactions? What combination of experiments must be performed to determine completely and uniquely the neutral-current structure?

Both approaches have shortcomings. In the first approach we are essentially assuming that half of the theory is correct and checking consistency by examining, for instance, whether $\sin^2\theta_W$ determined from $R_{\nu N}$ agrees with that determined from $R_{\bar{\nu} N}$.[44] In that way it is difficult to see whether models with qualitatively different sets of coupling parameters also fit the same data. In the second approach there are too many parameters to be determined, and short of good data, the rate of progress is expected to be very slow. Small wonder that the first approach is more popular in the experimental community; it is far easier to determine experimentally a single parameter than four (or more) parameters separately.

In this talk I follow the second approach. Much of what I am going to say in this section is contained in a series of papers I wrote in collaboration with Pham Quang Hung.[45] However, I would like to emphasize that many other people have worked along similar lines--Rajasekaran, Sarma, Ecker, Pietschmann, Bjorken[46,47]...to mention a few names. I am pleased to report today that even if we follow this more general approach, nontrivial advances have been made recently, particularly since last summer.

We first concentrate on the νN interactions. To the extent that the nucleon is made up of ordinary (u,d) quarks, it is difficult to determine the coupling parameters associated with strange and charmed quarks.[48] So, assuming V, A structure, we have four parameters,[49] (i) isovector (I=1) V, (ii) isovector A, (iii) isoscalar (I=0) V, and (iv) isoscalar A, where I = 0 stands for currents going like $\bar{u}u + \bar{d}d$ with no $\bar{s}s$, nor $\bar{c}c$. These parameters are denoted by α, β, γ, δ, respectively. Their precise definitions are given by the effective Lagrangian

$$L = -\frac{G}{\sqrt{2}}\,\bar{\nu}\gamma_\lambda(1+\gamma_5)\nu\,\Big\{\frac{1}{2}\,[\bar{u}\gamma_\lambda(\alpha+\beta\gamma_5)u - \bar{d}\gamma_\lambda(\alpha+\beta\gamma_5)d] + \frac{1}{2}\,[\bar{u}\gamma_\lambda(\gamma+\delta\gamma_5)u + \bar{d}\gamma_\lambda(\gamma+\delta\gamma_5)d]\Big\}\ . \tag{6.1}$$

My γ_5 is defined in such a way that $\alpha = \beta$, $\gamma = \delta$ for pure V minus A. Alternatively we may work with the "chiral coupling constants" for u and d defined by

$$\begin{aligned} \varepsilon_L(u) &= \frac{1}{4}\,(\alpha+\beta+\gamma+\delta)\ , & \varepsilon_L(d) &= \frac{1}{4}\,(-\alpha-\beta+\gamma+\delta)\ , \\ \varepsilon_R(u) &= \frac{1}{4}\,(\alpha-\beta+\gamma-\delta)\ , & \varepsilon_R(d) &= \frac{1}{4}\,(-\alpha+\beta+\gamma-\delta)\ . \end{aligned} \tag{6.2}$$

Let $\sigma_{NC}(\overset{(-)}{\nu}N)$ and $\sigma_{CC}(\overset{(-)}{\nu}N)$ be the cross sections (total or differential) for the inclusive reactions

$$\overset{(-)}{\nu} + N \to \overset{(-)}{\nu} + \text{hadrons}\ , \tag{6.3a}$$

$$\overset{(-)}{\nu} + N \to \mu^{-(+)} + \text{hadrons}\ , \tag{6.3b}$$

respectively, where N stands for the proton and the neutron averaged over. The key equations used to determine the coupling parameters are, first,[45,46]

$$[\sigma_{NC}(\nu N) + \sigma_{NC}(\bar{\nu}N)]/[\sigma_{CC}(\nu N) + \sigma_{CC}(\bar{\nu}N)]$$

$$= \frac{1}{4}(\alpha^2 + \beta^2 + \gamma^2 + \delta^2)$$

$$= |\varepsilon_L(u)|^2 + |\varepsilon_L(d)|^2 + |\varepsilon_R(u)|^2 + |\varepsilon_R(d)|^2 \,, \qquad (6.4a)$$

$$[\sigma_{NC}(\nu N) - \sigma_{NC}(\bar{\nu}N)]/[\sigma_{CC}(\nu N) - \sigma_{CC}(\bar{\nu}N)]$$

$$= \frac{1}{2}(\alpha\beta+\gamma\delta) = |\varepsilon_L(u)|^2 + |\varepsilon_L(d)|^2 - |\varepsilon_R(u)|^2 - |\varepsilon_R(d)|^2 \,. \qquad (6.4b)$$

The cross-section ratios given here measure, respectively, the overall strength and VA interference, but not isovector-isoscalar interference. Even though these relations were first obtained using the quark parton model, they are insensitive to most types of scaling violation suggested in the literature, e.g., the changing roles played by the valence and ocean quarks with increasing q^2. In fact (strong) isospin invariance, chiral symmetry--already tested in the charged-current reactions--and a very crude form of quark models [the kind of argument used in deriving the successful cross-section relation $\sigma(\rho N) = \sigma(\omega N)$] are sufficient for obtaining (6.4).[50]

The present status of the inclusive reactions on complex nuclei, studied by four different groups,[51] is summarized in Fig. 5. From this we conclude:

(i) Pure V + A is strongly excluded.

(ii) Pure V <u>or</u> pure A is also ruled out.

(iii) Pure V-A is unlikely.

So the neutral-current interactions appear to be neither chirality pure, nor parity pure.

As long as the inclusive reactions are studied with isoscalar targets (or targets with nearly the same number of protons and neutrons), we cannot separate out the isospin dependence of the hadronic neutral current. To study isoscalar-isovector interference we need inclusive data on n and p separately. Unfortunately data available so far on $R_{\nu p}$ and $R_{\bar{\nu}p}$ are too crude to yield useful information on the isospin dependence of the neutral-current couplings.

An alternative method for studying isoscalar-isovector interference ($\alpha\gamma+\beta\delta$ and $\alpha\delta+\beta\gamma$) is provided by the semi-inclusive pion production reactions

$$\overset{(-)}{\nu} + N \rightarrow \overset{(-)}{\nu} + \pi^{\pm,o} + \text{hadrons} \qquad (6.5)$$

where N again stands for n and p averaged over. If the current were isospin pure, i.e., pure isovector <u>or</u> pure isoscalar, we would expect for the pion charge ratios in (6.5)

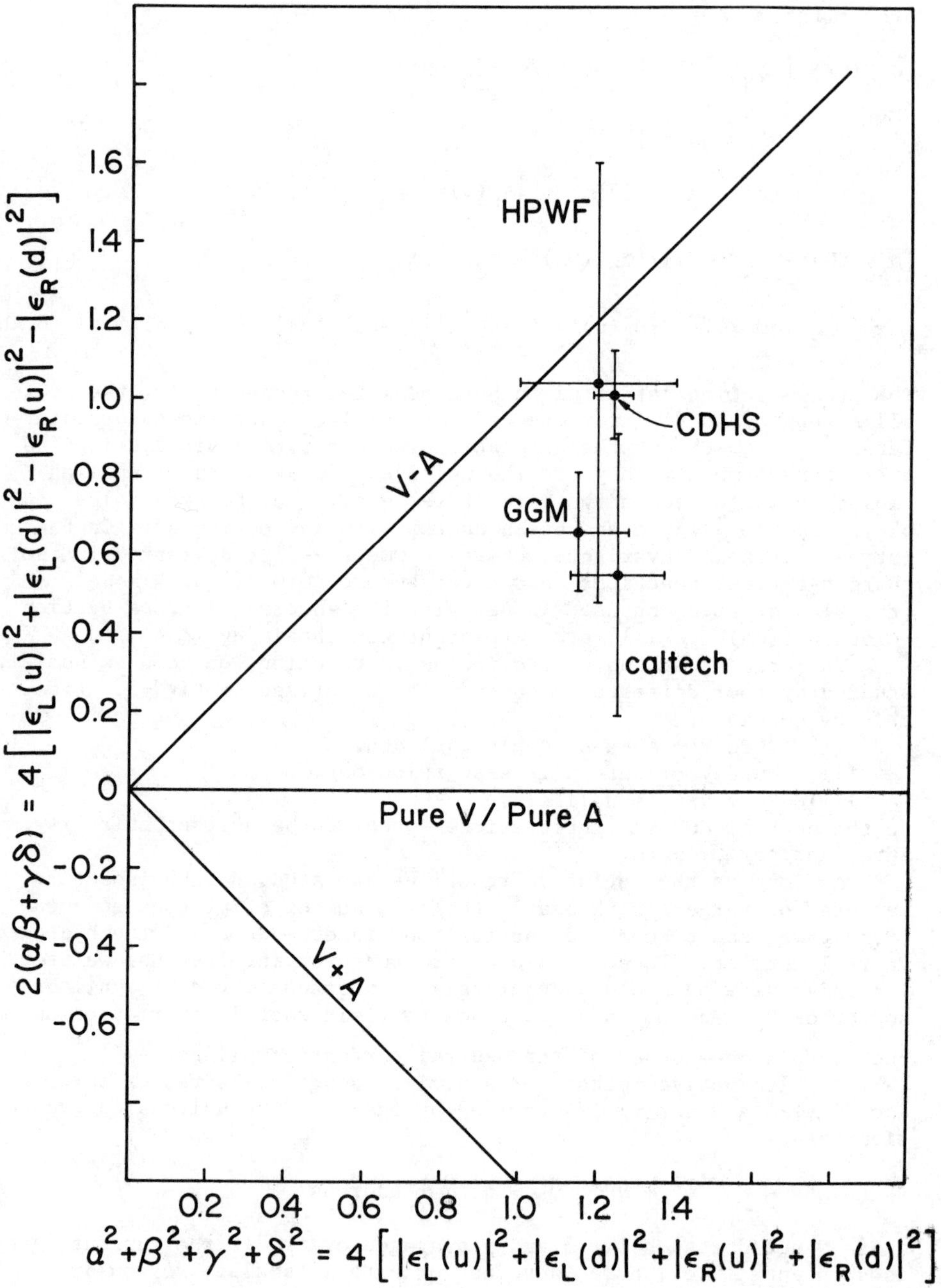

Fig. 5. Coupling constant determination from inclusive data.

$$(\pi^+/\pi^-)_{\nu\to\nu} = (\pi^+/\pi^-)_{\bar{\nu}\to\bar{\nu}} = 1 \quad . \tag{6.6}$$

Deviations of these ratios from unity would unequivocally prove that the current is not isospin pure.

For a quantitative extraction of the coupling parameters we must rely on the quark fragmentation model[52]; fortunately the method used has been successfully tested in electroproduction and charged-current reactions. To illustrate the basic physics involved, I just write down one of the eight key equations used[53]:

$$\frac{d\sigma}{dxdz}(\nu + p \to \nu + \pi^+ + \text{hadrons})$$
$$= \left\{ |\varepsilon_L(u)|^2 + \tfrac{1}{3}|\varepsilon_R(u)|^2 \right\} f_u^{(p)}(x)\, D_u^{\pi^+}(z) + [d \leftrightarrow u] \; . \tag{6.7}$$

Inside the curly bracket we have the chiral coupling constants defined earlier by (6.2); the bracket is multiplied by $f_u^{(p)}(x)$, the probability for finding the u quark inside the proton with fractional momentum x; next appears the fragmentation function (known as the "decay function" at Caltech) $D_u^{\pi^+}(z)$, which measures the probability for the u quark, struck by the current, to fragment into a π^+. The formula is applicable only in the quark fargmentation region, i.e., for events in which pions are produced in the direction of the momentum of the current with an appreciable value of z, the longitudinal momentum of the pion measured in units of its maximum value. The functions $f_u^{(p)}(x)$, $D_u^{\pi^+}(z)$, etc., are known from exhaustive studies of the electroproduction and charged-current reactions[54]; so the only unknown are the neutral-current coupling constants. The method works particularly well because $D_u^{\pi^+}/D_u^{\pi^-}$ is as large as three.

Experimentally, a subgroup of the Gargamelle Collaboration, located at Aachen,[55] obtained

$$\begin{aligned} (\pi^+/\pi^-)_{\nu\to\nu} &= 0.77 \pm 0.14 \; , \\ (\pi^+/\pi^-)_{\bar{\nu}\to\bar{\nu}} &= 1.64 \pm 0.36 \; , \end{aligned} \tag{6.8}$$

for $E_{\pi^\pm} > 1$ GeV, $0.3 < z < 0.7$. This shows the presence of isoscalar-isovector interference, independently of the details of any model.[56] Using these numbers together with the Gargamelle inclusive results, Sehgal,[57] a theorist at Aachen, succeeded in determining <u>uniquely</u> $|\varepsilon_L(u)|^2$, $|\varepsilon_L(d)|^2$, $|\varepsilon_R(u)|^2$, and $|\varepsilon_R(d)|^2$.

To start with, we had four constants to be determined; we now have four quantities measured. Yet, there are still ambiguities because the sign of each of the chiral coupling constants is unknown. Physically, this is not surprising in view of the incoherence assumption inherent in the quark parton model. In terms of α, β, γ, and δ we still have freedom in interchanging the roles of V and A and also the roles of isovector and isoscalar, separately for both the left-handed couplings (V-A type) and the right-handed couplings (V+A type).

The situation can be most easily visualized by displaying the allowed regions in left- and right-handed coupling planes.[45] See Fig. 6. From the inclusive reactions the allowed values of the coupling constants must lie inside the annular regions of Fig. 6. When the inclusive data are combined with the semi-inclusive pion data, only the interiors of the small oval regions are allowed. Note reflection invariance about any of the chiral axes, which illustrates the sign ambiguities ($V \leftrightarrow A$ and $I = 0 \leftrightarrow I = 1$ ambiguities) referred to earlier.

We can construct various coupling-constant solutions by pairing allowed regions in the left-handed coupling plane with ones in the right-handed coupling plane. The solutions obtained are labeled as A, B, C, D, A', B', C', and D', as shown. A primed solution is related to the corresponding unprimed solution by changing the sign of each coupling constant simultaneously. Since the overall sign is unmeasurable except by looking at interference with gravity, we consider only the unprimed solutions.

The four solutions obtained are displayed in Table 1.[45] It is seen that Solutions A and B are isovector dominant while Solutions C and D are isoscalar dominant. Notice that the four solutions are related to each other by $V \leftrightarrow A$ and/or $I = 1 \leftrightarrow I = 0$.

	α (I=1, vector)	β (I=1, axial)	γ (I=0, vector)	δ (I=0, axial)
Solution A	0.45±0.14	0.92±0.14	-0.35±0.15	0.12±0.15
Solution B	0.92±0.14	0.45±0.14	0.12±0.15	-0.35±0.15
Solution C	-0.12±0.15	0.35±0.15	-0.92±0.14	-0.45±0.14
Solution D	0.35±0.15	-0.12±0.15	-0.45±0.14	-0.92±0.14
SW($\sin^2\theta_W$ = 0.25)	0.5	1	-0.17	0
SW($\sin^2\theta_W$ = 0.30)	0.4	1	-0.2	0
SW($\sin^2\theta_W$ = 0.35)	0.3	1	-0.23	0

Table 1.

Phenomenological solutions in the neutrino-hadron interactions.

To obtain further constraints on the coupling parameters we turn to exclusive reactions. First, a detailed analysis of elastic scattering[45,58]

$$\overset{(-)}{\nu} + p \to \overset{(-)}{\nu} + p \ , \tag{6.9}$$

which I don't go into here shows that the isoscalar dominant solutions, C and D, are ruled out by the data of the HPW Collaboration.[59] An additional, even stronger piece of evidence against C and D comes from the recent observation by the Gargamelle propane Collaboration[60] of a very clean and strong Δ signal in single pion production

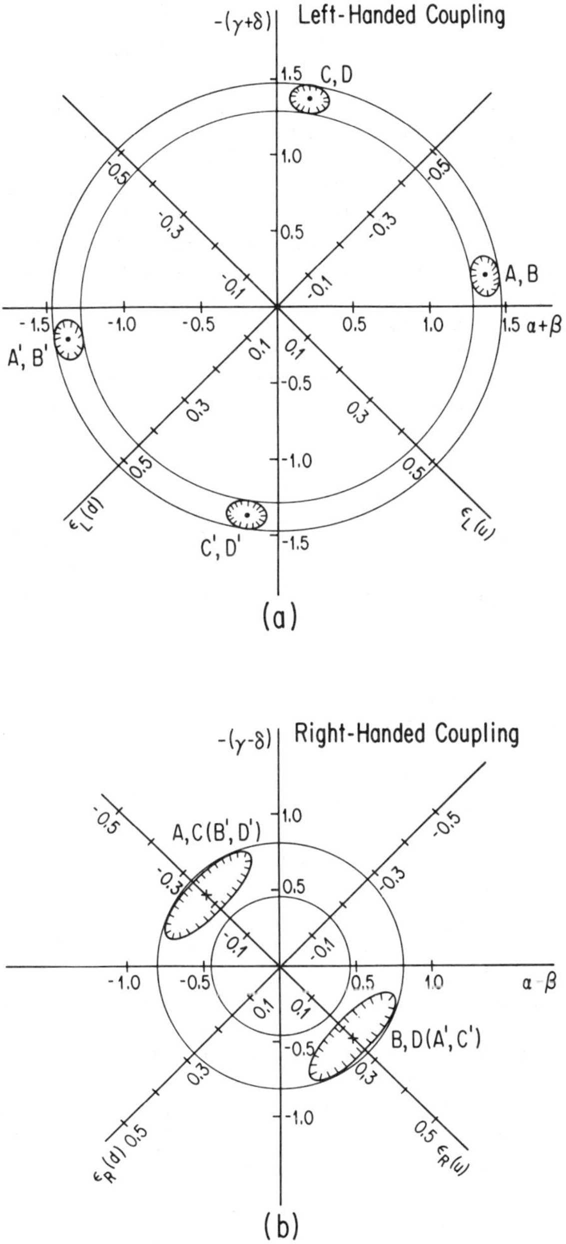

Fig. 6. Allowed regions in coupling-constant planes.

$$\overset{(-)}{\nu} + p \rightarrow \overset{(-)}{\nu} + \Delta^+(1238) \qquad \Delta^+ \rightarrow p + \pi^0 \qquad . \qquad (6.10)$$

See Fig. 7, which also shows the $\pi^- p$ mass distribution. Notice that Δ^+ shows up in the neutral-current reaction even more clearly than in the analogous charged-current reaction.

To sum up we have an "almost unique" determination of the coupling constants in the neutrino-hadron sector. The two surviving solutions, Solutions A and B, are both isovector dominant, but not pure isovector, and are related to each other by interchanging the roles of V and A.

To appreciate the significance of the entries in Table 1, suppose the charged-current and neutral-current interactions were SU(2) symmetric, i.e., related by weak-isospin rotations, with left-handed doublets only.[14] We would then expect, with our normalization convention,

$$\alpha = \beta = 1 \ , \qquad \gamma = \delta = 0 \quad . \qquad (6.11)$$

It is seen that none of the four solutions is compatible with (6.11). On the other hand, if the hadronic part of the neutral currents were proportional to the electromagnetic current,[61] we would expect

$$\alpha = 3\gamma = \rho \ , \qquad \beta = \delta = 0 \quad , \qquad (6.12)$$

which also disagrees with all four solutions. However, suppose we form a linear combination of the SU(2) current and the electromagnetic current:

$$J_\lambda^{NC} = J_\lambda^{\ 3} - \text{"const"}\ J_\lambda^{e.m.} \qquad . \qquad (6.13)$$

We see that Solution A is in good agreement with (6.13) provided the "constant" is taken to be ~ 0.3. Note that (6.13) is just of the form required by the "standard model" (denoted as SW in Table 1) à la Glashow,[17] Salam,[25] Ward,[18] and Weinberg[24] with the "constant" identified with $\sin^2\theta_W$. So even though our approach has been model-indendent throughout, one of the two surviving solutions is seen to agree with the predictions of the standard model. Bjorken[62] estimates that in a completely random model satisfying rough order-of-magnitude universality between the charged and neutral currents, the probability for all four coupling constants coinciding with the standard model predictions with the kind of errors indicated in Table 1 is a few % or less.

VII. DETERMINATION OF NEUTRAL-CURRENT COUPLINGS-- THE ELECTRONIC CURRENT

Let us now turn to the electronic current ($\bar{e}e$), which can be studied in $\nu_\mu e$, $\bar{\nu}_\mu e$, and $\bar{\nu}_e e$ scattering, (5.3d) and (5.3e). Our goal here is to determine the coupling constants g_V and g_A defined by

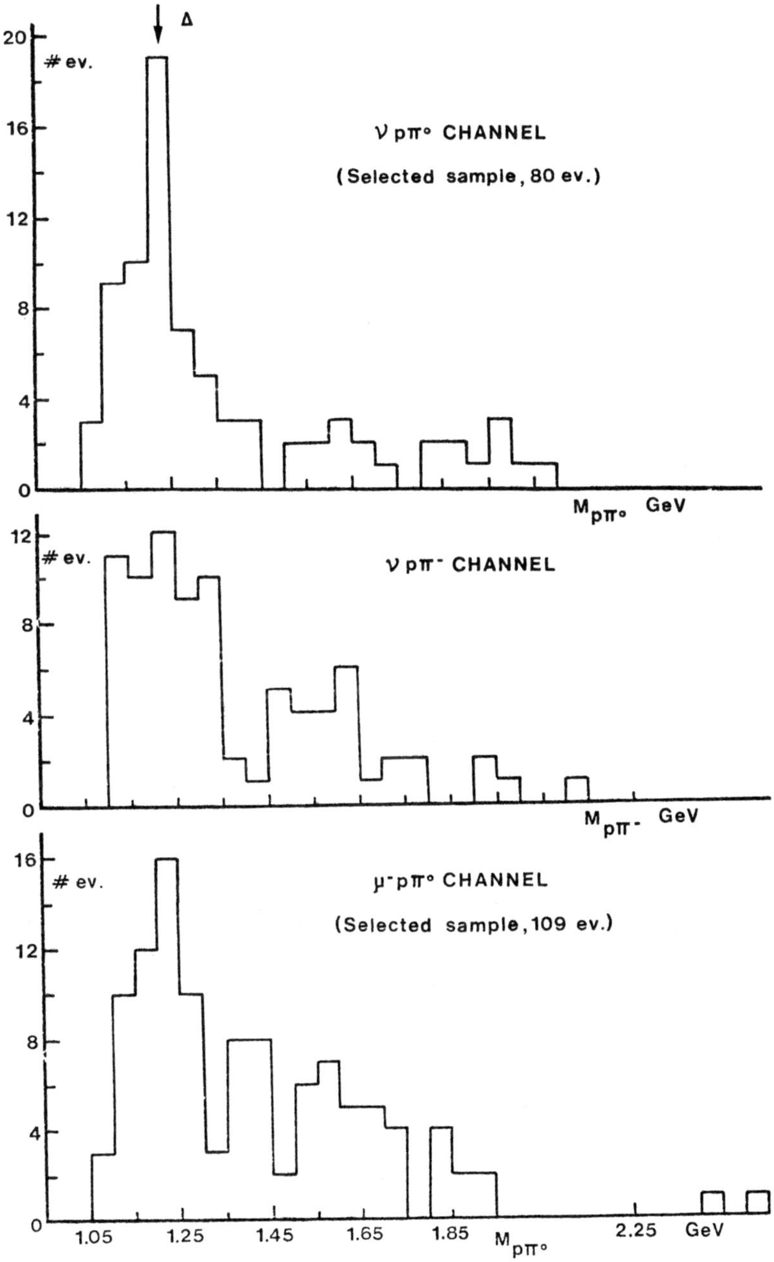

Fig. 7. The mass-distributions of the $\pi^o p$ $(\pi^- p)$ system in νN collisions.

$$L = -\frac{G}{\sqrt{2}}\bar{\nu}_\mu\gamma_\lambda(1+\gamma_5)\nu_\mu\,\bar{e}(g_V\gamma_\lambda + g_A\gamma_\lambda\gamma_5)e \qquad (7.1)$$

This subject has been reviewed recently[63]; so I confine myself to a few remarks only.

First, we note that a completely model-independent analysis is rather difficult here. The reason is that there is again a vector-axial-vector ambiguity ($g_V \leftrightarrow g_A$), which, this time, can be resolved only by looking at a term of order m_e/E_ν. For example, suppose the data showed $\sigma(\nu_\mu e) = \sigma(\bar{\nu}_\mu e)$. This would indicate that the ($\bar{e}e$) current is pure vector <u>or</u> pure axial-vector but, for all practical purposes, no accelerator neutrino experiment could be performed to decide which of the two possibilities is realized. The only hope may lie in super low-energy reactor experiments which may barely detect the elusive m_e/E_ν term.

The most model-independent way of displaying the data is again to sketch the allowed regions in a coupling-constant plane.[64] See Fig. 8 where the $\bar{\nu}_e e$ data of the Irvine Group[65] (working at Savannah River) and the Gargamelle $\bar{\nu}_\mu e$ and $\nu_\mu e$ results[66] are displayed. In drawing the ellipse for $\bar{\nu}_e e$ scattering μe universality has been assumed.

Turning now to the model-dependent approach, we note that the standard model predicts

$$g_V = -\frac{1}{2} + 2\sin^2\theta_W\,, \qquad g_A = -\frac{1}{2}\ . \qquad (7.2)$$

If we analyze the three elastic reactions <u>assuming</u> $g_A = -1/2$, then the data shown in Fig. 8 are seen to be consistent with $\sin^2\theta_W$ of order 0.3. However, there is a more recent work of an Aachen-Padua group who claims that $\sigma(\nu_\mu e) = \sigma(\bar{\nu}_\mu e)$ is ruled out especially when the differential distributions are studied in detail; this means that $\sin^2\theta_W$ <u>cannot</u> be around 0.25 (pure axial-vector).[67]

Parity violation in bismuth atoms explored by linearly polarized laser beams is another important (and controversial!) place where the electronic current has been studied. The existence of a parity-violating potential between the electron and the nucleon due to weak neutral currents would imply that atomic levels are not pure parity eigenstates. As a result, a predominantly M1 transition between a pair of atomic levels--say, between $^2D_{3/2}$ and $^4S_{3/2}$ of the $6p^3$ configuration of Bi--is expected to contain a tiny E1 component, of order 10^{-7} in the amplitude. This will cause a rotation of the polarization plane of a laser beam as the beam goes through bismuth vapor.[68] The effect is sensitive to interactions of the form $A_{electron}V_{hadron}$, or, more quantitatively to

$$Q_W = -[\tilde{\alpha}(Z-N) + 3\tilde{\gamma}(Z+N)] \qquad (7.3)$$

where $\tilde{\alpha}$ and $\tilde{\gamma}$ are defined in a manner similar to α and γ of (6.1)

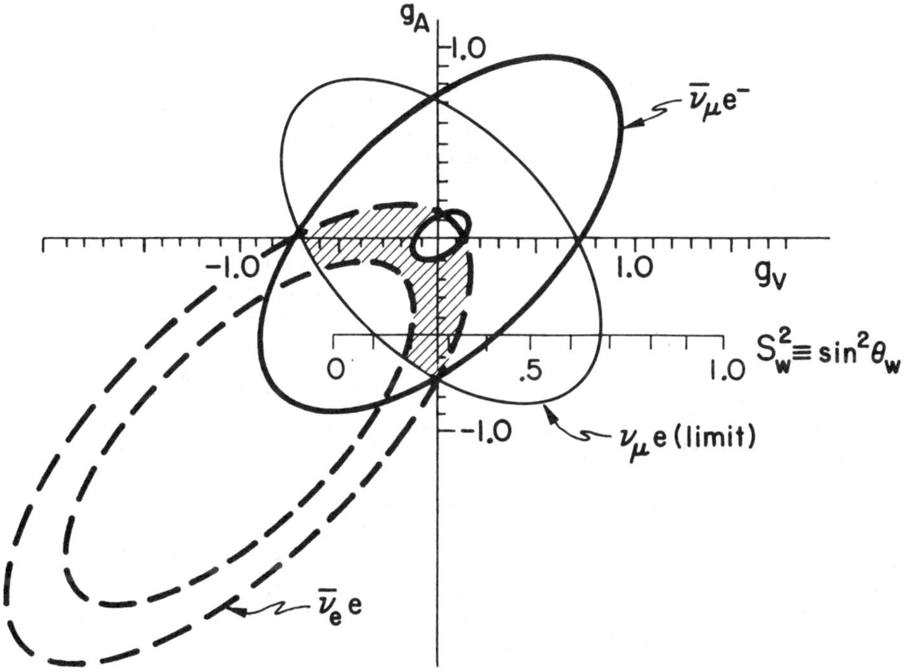

Fig. 8. Determination of the νe neutrino-electron scattering constants, g_V and g_A.

with $\bar{\nu}\gamma_\lambda(1+\gamma_5)\nu$ replaced by $\bar{e}\gamma_\lambda\gamma_5 e$, and Z = number of protons, N = number of neutrons.

As of last summer, there were two parity experiments--Seattle[69] and Oxford[70]--which showed essentially null results; the upper limits for optical rotation, if any, were reported by both groups to be substantially lower than the standard-model predictions despite some uncertainties in atomic-physics calculations.[71] There has been, however, a new development in the past several weeks; I received a telegram from Barkov and Zolotoryov[72] of Novosibirsk reporting that optical rotation in bismuth vapor has been observed, four standard deviations away from parity conservation, compatible with the standard model prediction. Clearly confirmation of this important result is needed. In any case, it is gratifying from the point of view of unity of science, that some of the questions raised by performing accelerator experiments with neutrino beams of 2 to 200 GeV are now being answered by performing experiments with laser beams of $\sim 10^{-9}$ GeV.

VIII. COMPARISON WITH GAUGE MODELS

Since this session is supposed to be concerned with "Applications of Gauge Theories," let me now proceed to compare the experimental data with gauge models. As is well known, besides the standard model, there have been a number of unified gauge models of weak and electromagnetic interactions proposed in the last several years. That we have so many models is not surprising; it is rather easy to construct a gauge model. I can even provide a form for writing an original paper if you are interested in model building. See Fig. 9.

The standard model is the simplest $SU(2) \otimes U(1)$ model where all left-handed fermions are in weak isospin doublets while all right-handed fermions appear as singlets. We may still stay within the same group, $SU(2) \otimes U(1)$, and introduce right-handed doublets, $(u,b)_R$ and/or $(t,d)_R$. Alternatively we may enlarge the underlying group to consider $SU(3) \otimes U(1)$, $SU(2) \otimes SU(2) \otimes U(1)$, etc. Some examples of gauge models are shown in Table 2. I would like to emphasize that this is only a partial list of the gauge models proposed.

Let us see how the various models confront the neutral-current data. We eliminate models, one by one, by putting x in Table 2 as we go along. The results of the inclusive reactions (6.3a), as summarized in Fig. 5, eliminate a pure SU(2) V-A model[14,73]; also ruled out is the simplest form of vector-like models based on $SU(2) \otimes U(1)$ where the right-handed doublets, $(u,b)_R$ and $(t,d)_R$, both appear with

Form for writing an "original" paper on gauge models

Theorist J. J. Sakurai of UCLA recently offered a form that should simplify the writing of papers on gauge models. At the Conference on Leptons and Quarks held at Irvine in December, he presented the following form:

We present a new gauge theory of weak and electromagnetic interactions based on the group...

Under (   ) the left- and right-handed quark multiplets transform as...

We now specify the Higgs sector...

The mass formulas for the gauge bosons can be readily obtained as follows...

Triangular anomalies are cancelled by postulating additional heavy leptons...

The inclusive νN and $\bar{\nu} N$ cross sections calculated using the interaction (   ) agree with the predictions of the Weinberg–Salam model if we identify...

Trimuon events are possible in our model because...

The bismuth puzzle is resolved by considering...

We wish to thank the Aspen Center for Physics for its hospitality. This work was supported in part by...

1. S. Weinberg, Phys. Rev. Lett. **19**, 1264 (1967); A. Salam in "Elementary Particle Physics" (ed. N. Svartholm, Stockholm 1968), p. 367.

20 PHYSICS TODAY / APRIL 1978

Fig. 9. Form for writing an original paper on gauge models.

Inclusive	π - semi-inclusive	$\overset{(-)}{\nu}p$ elastic, single π	atoms (Seattle, Oxford)	atoms (Novosibirsk)	Models		
x	x	x		x	SU(2)[14,73]	$(u,d)_L$, $(\nu_e,e^-)_L$ ...	isovector V-A
	A		x		SU(2) × U(1)[17,18,24,25]	$(u,d)_L$, $(\nu_e,e^-)_L$, u_R, d_R, e_R..	"standard"
x	x	x	x		SU(2) × U(1)[74]	$(u,d)_L$, $(u,b)_R$, $(t,d)_R$ ...	"vector-like"
	B		(x)	x	SU(2) × U(1)[76]	$(u,d)_L$, $(u,b)_R$, d_R ...	
	x		x	x	SU(2) × U(1)	$(u,d)_L$, $(t,d)_R$, u_R ...	
	A			x	SU(2) × U(1)[79]	$(\nu_e,e^-)_L$, $(E^o,e^-)_R$; quarks standard	
	x			x	SU(3) × U(1)[77]	$(u,d,b)_L$, $(t,b,d)_R$, $(\nu_e,e^-,E^-)_L$, $(E^o,E^-,e^-)_R$	
	B			x	SU(3) × U(1)[80]	$(u,d,b)_L$, $(u,b,d)_R$, $(e^-,\nu_e,E_1{}^o)_L$, $(e^-,E_1{}^o,E_2{}^o)_R$	
	A			x	SU(2) × SU(2) × U(1)[81]	parity conservation with Z_V and Z_A	

Table 2. Confrontation of gauge models with experimental data.

full strength together with the familiar left-handed doublets $(u,d_C)_L$ and $(c,s_C)_L$.[74] Further evidence against this version of vector-like models comes from exclusive reactions, elastic $\overset{(-)}{\nu}p$ scattering[59] and Δ production,[75] where, as in the inclusive reactions, strong VA interference has been observed. The semi-inclusive pion reactions (6.5) rule out all SU(2)⊗U(1) models where the right-handed doublet $(t,d)_R$ appears with appreciable strength; an SU(3)⊗U(1) model with $(t,b,d)_R$[77] is also in difficulty. The surviving models are indicated by A or B in Table 2 according to whether Solution A or B fits the model in question.

There is a nontrivial consistency check between $\overset{(-)}{\nu}e$ scattering and the neutrino-hadron interactions that must be satisfied in a wide class of SU(2)⊗U(1) models. Almost all model builders working with SU(2)⊗U(1) keep the following assignment for the left-handed fermions

$$T^{\nu}_{3L} = T^{u}_{3L} = 1/2 \;, \qquad\qquad (8.1)$$

$$T^{e}_{3L} = T^{\mu}_{3L} = T^{d}_{3L} = -\,1/2$$

to insure charged-current universality. Bernabéu and Jarlskog[78] have noted that in SU(2)⊗U(1) models satisfying (8.1) with <u>any</u> T_{3R} assignment, there is a linear relation between the $\overset{(-)}{\nu}e$ scattering coupling constants and the hadronic coupling constants:

$$\alpha + \beta + 3(\gamma+\delta) + 2(g_V + g_A) = 0 \ . \tag{8.2}$$

This relation is beautifully satisfied by the two surviving solutions in the neutrino-hadron sector, Solutions A and B. They also derive, under the same assumption on T_{3L} and T_{3R},

$$\alpha + \beta - 3(\gamma+\delta) - 2\rho = 0 \ , \tag{8.3}$$

where ρ characterizes the overall strength of the neutral-current interactions relative to the charged-current interactions. In models where a single scalar doublet is responsible for symmetry breakdown à la Higgs, ρ must be unity; in more general models we expect

$$\rho = m_W^{\ 2}/m_Z^{\ 2} \cos^2\theta_W \ . \tag{8.4}$$

Experimentally we obtain, for both Solutions A and B,

$$\rho = 1.02 \begin{smallmatrix} +\ 0.03 \\ -\ 0.18 \end{smallmatrix} \ , \tag{8.5}$$

in remarkable agreement with $\rho = 1$. So if SBGT based on SU(2)⊗U(1) is on the right track, then we must have the simplest Higgs mechanism of the kind discussed in Weinberg's 1967 paper.[24]

The final experimental constraints we must examine are the implications of the atomic parity-violation (or parity-conservation?) experiments. Suppose the parity experiments reported last summer[69,70] turned out to be correct and the atomic-physics calculations used are proved to be reliable. The standard model would then be ruled out; an SU(2)⊗U(1) model where $(u,b)_R$ enters with somewhat reduced strength could still survive. The null bismuth results would be consistent with models incorporating the right-handed lepton doublet $(E^0,e^-)_R$, which makes the electronic current pure vector,[79] with models based on SU(3)⊗U(1) invented to give no parity violation in heavy atoms[77,80] (no $A_{\text{lepton}}V_{\text{hadron}}$ terms), and also with a class of SU(2)⊗SU(2)⊗U(1) models with two Z bosons mediating parity-conserving VV and AA interactions.[81]

On the other hand, if future experiments confirm the new Novosibirsk result, among all the gauge models proposed so far, the standard model--by far the simplest model incorporating unification --will be the sole survivor.

IX. EXPERIMENTAL TASKS FOR THE IMMEDIATE FUTURE

It is obviously desirable to have better data on reactions like $\nu_\mu e$ scattering, $\bar{\nu}_\mu e$ scattering, and $\overset{(-)}{\nu}p$ elastic scattering. Of particular importance is another independent attempt to extract

$|\varepsilon_{L,R}(u)|^2$ and $|\varepsilon_{L,R}(d)|^2$ from the semi-inclusive reactions (6.5) carried out at higher energies where the quark fragmentations model is believed to be more reliable. Alternatively, high-statistics experiments on the target dependence (p vs. n) of the inclusive reactions[45,46] would be very useful.

We have seen earlier that, if we have been on the right track in our phenomenological analysis of the neutrino-hadron interactions, only two possibilities now remain. The choice to be made is between Solution A and Solution B of Table 1. The last remaining ambiguity can be most clearly resolved by studying

$$\bar{\nu}_e + D \rightarrow \bar{\nu}_e + n + p \tag{9.1}$$

where D stands for the deuteron (not the recently discovered D meson). At reactor energies this reaction is allowed only through the isovector axial-vector piece of the hadronic neutral current,[82] and the cross section is therefore directly proportional to β^2, which is about four times larger for Solution A than for Solution B. An experiment to study (9.1) is currently in progress by Reines and collaborators at Savannah River.[83]

There are other means for distinguishing Solution A from Solution B--diffractive A_1 and ρ production, nuclear level excitation by intermediate-energy neutrinos,[84] polarization measurements in Δ production[85] and in elastic νp scattering,[86] etc.--but I won't discuss them in detail.

While the written version of this talk was being prepared, I received papers on single-pion production by Abbott and Barnett[87] and by Monsay.[88] These authors use Adler's pion production model[89] which includes a careful treatment of nonresonant multipoles, as well as Δ, to study the constraints imposed by recent pion production data. Of particular interest is the quantity

$$R_o \equiv \frac{\sigma(\nu n \rightarrow \nu n \pi^o) + \sigma(\nu p \rightarrow \nu p \pi^o)}{2\sigma(\nu n \rightarrow \mu^- p \pi^o)} \quad . \tag{9.2}$$

This ratio is calculated for the conditions of the Gargamelle propane experiment to be 0.33 for Solution A and 0.17 for Solution B while the experimental value is 0.45 ± 0.08.[90] So Solution B is not favored. An even more serious discrepancy with Solution B is encountered for $\bar{R}_o$, the neutral-to-charged current ratio for π^0 production by antineutrinos defined in a similar manner. It is to be noted, however, that earlier data (both bubble chamber[91] and counter[92]) gave values of R_o considerably lower--by a factor of about two--than the new Gargamelle value. In addition there are uncertainties associated with nuclear charge-exchange corrections and with the Adler model itself. For these reasons I prefer to wait and reserve judgment on this matter.

Atomic parity experiments may be performed in hydrogen and deuterium[93]; even though such experiments are more formidable, the interpretations are free from wave-function uncertainties. The same information can be obtained at high energies by performing inelastic scattering of polarized electrons on protons and deuterons[94]; what one must measure here is the difference between $\sigma(e_L p)$ and $\sigma(e_R p)$, or between $\sigma(e_L D)$ and $\sigma(e_R D)$, to a few parts in 10^5 at $E \simeq 20$ GeV, $q^2 \simeq 1$ GeV^2. An experiment along this line is in progress at SLAC by the Taylor group.

Weak-electromagnetic interference can be studied in the pure leptonic reaction

$$e^+ + e^- \to \mu^+ + \mu^- \tag{9.3}$$

at PETRA and PEP. Crude arguments based on dimensions show that the ratio of the weak to electromagnetic amplitude must go roughly as

$$G/(e^2/s) \simeq 10^{-4}\ s/\text{GeV}^2 \quad , \tag{9.4}$$

which is of order 10% at typical PETRA or PEP energies ($s \simeq 10^3$ GeV^2). The presence of axial-vector in the leptonic neutral current can be established by detecting a forward-backward asymmetry ($\cos\theta$ term) in the angular distribution of (9.3).[21,95] The magnitude of asymmetry expected is $\sim$ 8% in the standard model at $\sqrt{s}$ = 30 GeV. An experiment to detect such an interference effect is being planned by an MIT-DESY-Peking collaboration at PETRA.

I earnestly hope that by the time these proceedings appear in print, one or two of the experiments mentioned in this section will have been completed with decisive results.

X. ALTERNATIVES TO UNIFIED GAUGE THEORIES

It is difficult to overestimate the aesthetic appeal of unified gauge theories of electromagnetic and weak interactions. This is evidenced by the large number of bright theorists assiduously working in this field. Furthermore, we have seen that the simplest version of such unified models, the standard model à la Glashow, Salam, Ward, and Weinberg, has already scored nontrivial triumphs when confronted with the current experimental data. Even so, as in any scientific endeavor, we must be objective and unprejudiced. To this end we now examine if viable alternatives to unified gauge models can still be constructed. In contrast to the large number of theorists working on unified gauge theories, there are only three people in the world engaged in the heresy of contemplating "alternatives"--Bjorken[62] at SLAC, my collaborator, Pham Quang Hung, and myself.[96]

To be definite, let us suppose that the entire low-energy phenomenology of the neutral current-interactions--including atomic parity violation--is in perfect agreement with the standard model. One may then naturally ask: Do the successes of the standard model at low q^2 (i.e., $q^2 \ll m_W^2, m_Z^2$) necessarily prove the correctness of SBGT? Is it still possible to construct a viable alternative model that gives substantially different predictions at high energies?

Imagine history were different, and the neutral-current structure at low q^2 had been known experimentally prior to the work of Salam, Weinberg, etc. We would then have observed the following. There is a simple prescription for getting the right interaction; just let

$$J_\lambda^{\ 3} \rightarrow J_\lambda^{\ 3} - \text{"}\sin^2\theta_W\text{"}\ J_\lambda^{e.m.} \tag{10.1}$$

in the SU(2) symmetric current-current interaction (à la Bludman[14])

$$L = \frac{4}{\sqrt{2}}\, G\, \underset{\sim}{J}_\lambda \cdot \underset{\sim}{J}_\lambda \tag{10.2}$$

presumably generated by $W^{\pm,0}$ exchange. This suggests some kind of γW^0 mixing, to which we now turn our attention.

First, let us briefly review the formalism of two-particle mixing developed more than a decade ago.[97] For a single particle we have the propagator $(q^2+m^2)^{-1}$, and the mass is determined by looking at the zero of the inverse propagator $[(q^2+m^2) = 0]$. To generalize this to the two-particle case, we consider a two-by-two propagator matrix $\Delta(q)$ whose inverse $\Delta^{-1}(q)$ depends linearly on q^2 in the pole approximation. The masses of the two particles are determined by the roots of the quadratic equation

$$\det[\Delta^{-1}(q)]\Big|_{q^2=-m_1^{\ 2},-m_2^{\ 2}} = 0 \ . \tag{10.3}$$

Specializing now to our case, we assume, to start with:

(i) In addition to the photon there is a triplet of spin-one bosons $W^{\pm,0}$, all having the same mass m_W.

(ii) The photon is massless and coupled to the usual electromagnetic current.

(iii) The couplings of the W triplet to the weak-isospin current satisfy SU(2).

We now switch on an interaction that mixes γ and W^0; from a more fundamental point of view this may be an effective interaction through quark-pair and lepton-pair loops. We assume that this mixing interaction is the only mechanism that breaks SU(2). The γW^0 complex is characterized by the two-by-two inverse propagator matrix

$$\Delta^{-1}(q) = \begin{pmatrix} q^2 & \lambda q^2 \\ \lambda q^2 & q^2+m_W^{\ 2} \end{pmatrix} \ . \tag{10.4}$$

The off-diagonal elements are required to go like q^2 (i.e., no q^2 independent, constant term) if, after mixing, the photon is to remain massless and initially neutral particles like neutrinos are not to acquire electric charges. The masses of the physical particles after mixing can now be determined by looking at (10.3):

$$m_1^2 \equiv m_\gamma^2 = 0 \quad , \tag{10.5a}$$

$$m_2^2 \equiv m_Z^2 = m_W^2/(1-\lambda^2) \quad . \tag{10.5b}$$

To obtain the basic interaction due to the propagation of γ and Z we consider

$$L_{\text{eff}} = \frac{1}{2} J^\dagger \Delta J \tag{10.6}$$

with

$$J \equiv \begin{pmatrix} eJ_\lambda^{em} \\ gJ_\lambda^{3} \end{pmatrix} \tag{10.7}$$

where g is the usual dimensionless weak-interaction coupling constant for the $W^\pm$ boson related to Fermi's G by

$$G/\sqrt{2} = g^2/8m_W^2 \quad . \tag{10.8}$$

Inverting (10.4) and expressing everything in terms of m_Z^2, we obtain the final result

$$\frac{1}{2} J^\dagger \Delta J = \frac{1}{2} \{ e^2 J_\lambda^{em} \frac{1}{q^2} J_\lambda^{em}$$

$$+ \frac{g^2}{(1-\lambda^2)} [J_\lambda^3 - (\lambda e/g) J_\lambda^{em}] \frac{1}{q^2+m_Z^2} [J_\lambda^3 - (\lambda e/g) J_\lambda^{em}] \} \quad . \tag{10.9}$$

Notice that the neutral boson mass has increased [see (10.5b)] but the coupling constant has also increased by the same factor. As a result, the second term in (10.9), as $q^2 \to 0$, is just

$$\frac{g^2}{m_W^2} [J_\lambda^3 - (\lambda e/g) J_\lambda^{em}] [J_\lambda^3 - (\lambda e/g) J_\lambda^{em}] \quad . \tag{10.10}$$

This is recognized to be precisely the neutral-current interaction of the standard model at low q^2 provided we identify

$$\lambda e/g = \sin^2\theta_W \quad . \tag{10.11}$$

Even the overall strength is the same as in spontaneously broken $SU(2) \otimes U(1)$ models with the simplest Higgs mechanism. To sum up, all the successful features of the standard model at low energies can be reproduced just as well in this very simple model based on γW^0 mixing.

Does this alternative model give predictions very different from those of the standard model at high energies? The answer is affirmative: the model can accommodate values of the weak boson masses very different from Weinberg's. In the standard model a knowledge of $\sin^2\theta_W$, determined from low energy experiments, is sufficient to give <u>both</u> the W and the Z mass [see (4.2)]. On the other hand, in the simple γ-W^0 mixing model we have weaker predictions

$$m_Z^2 = m_W^2/[1 - (m_W\text{"}\sin^2\theta_W\text{"}/37.4\ \text{GeV})^2] \quad , \tag{10.12a}$$

$$m_W < 37.4\ \text{GeV}/\text{"}\sin^2\theta_W\text{"} \quad . \tag{10.12b}$$

A knowledge of what the experimentalists call "$\sin^2\theta_W$" is insufficient to determine the W and the Z mass separately. If we know either one of them, then the other will be predicted, as shown in Fig. 10. This alternative model cannot tell us whether the Z boson is at 25 GeV or 200 GeV, but, if, for instance, experiments at CERN ISR discover the Z boson at 25 GeV, then the W mass is predicted to be 24.5 GeV. Notice also that we get an upper limit on the W mass.

The true believer of the standard model will say, "The spectacular successes of the standard model in accounting for low-energy neutral-current data show that we are unmistakably on the right track. The W and Z are expected at the masses predicted by Weinberg with certainty:

$$\begin{aligned} m_W &= (65 - 75)\ \text{GeV} \\ m_Z &= (80 - 95)\ \text{GeV.''} \end{aligned} \tag{10.13}$$

The heretic will say, "A viable alternative is still possible. It is worth performing experiments at high-energy colliding-beam facilities to see if Weinberg is really right."

XI. EXPERIMENTAL SEARCH FOR W AND Z

Many high-energy physicists believe that the central experimental task of elementary particle physics of the '80s and the '90s is to discover the carriers of weak forces, $W^{\pm}$ and Z , and study their properties. Colliding-beam facilities are being planned in various parts of the world with just that goal in mind.

Consider the production of W or Z in proton-proton collisions:

$$p + p \rightarrow W\ (\text{or } Z) + \text{any} \quad . \tag{11.1}$$

The cross section for this process can be estimated by appealing to Drell-Yan scaling[98]

$$\frac{d\sigma}{dm^2}\,(p + p \rightarrow e^+ + e^- + \text{any}) = \frac{1}{m^4}\,F(m^2/s) \quad , \tag{11.2}$$

(where m stands for the invariant electron-pair mass) CVC[99] and

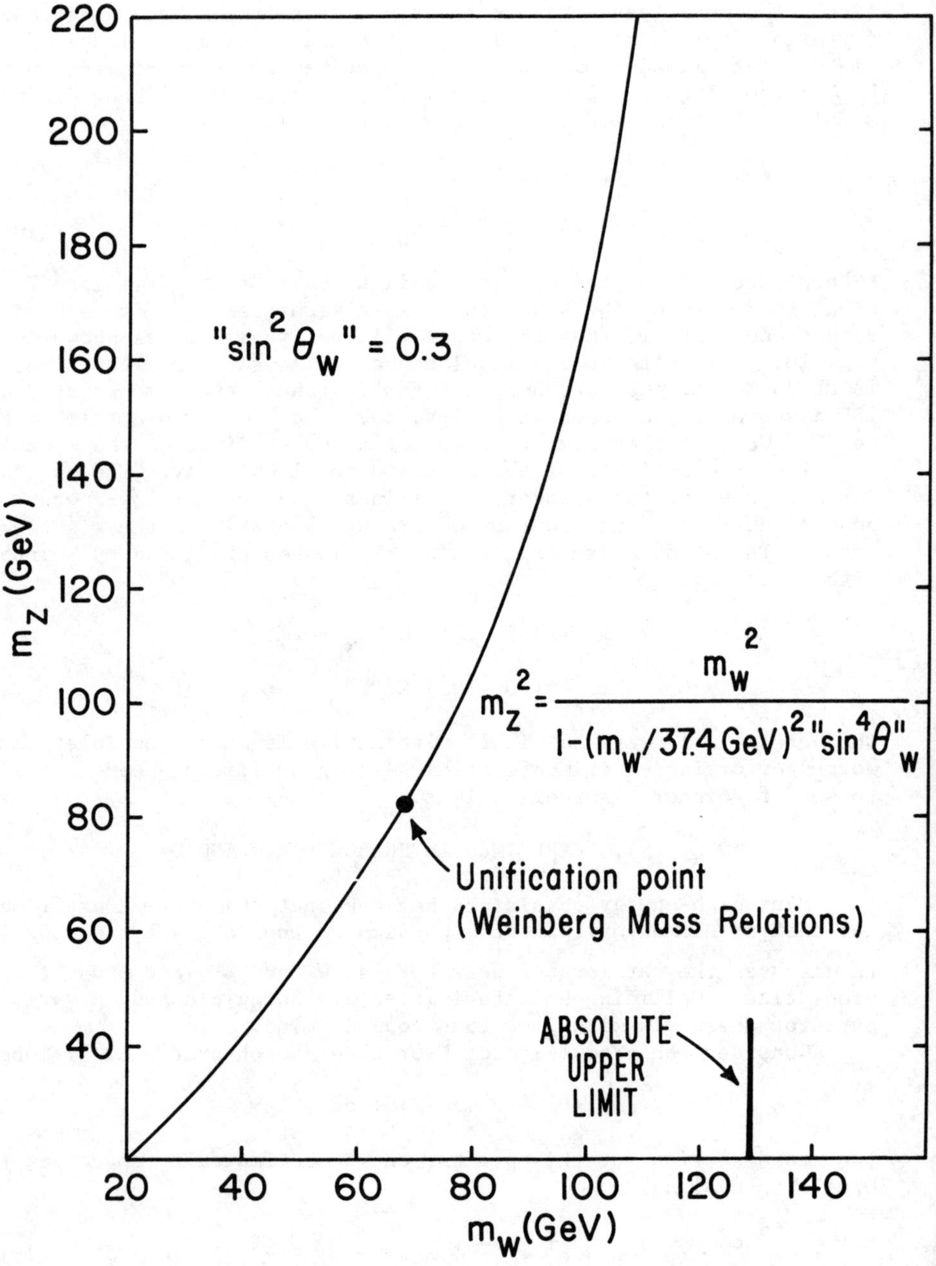

Fig. 10. Boson mass prediction in the γ-W^0 mixing model.

chiral symmetry, or the quark-parton model that incorporates all these ideas. Estimates for W and Z production have been made by various authors--there is even a lengthy review paper by Quigg[100] devoted just to this subject--and it is generally agreed that for $\sqrt{s} \sim 400$ GeV and m_W and m_Z in the Weinberg range (70-90 GeV), colliding beam luminosities of $\sim 10^{31}$ cm^{-2} sec^{-1} are comfortably sufficient. However, in case such theoretical estimates are wrong by as much as two orders of magnitude it is safer to plan a facility with luminosities $\sim 10^{33}$ cm^{-2} sec^{-1}, which are, in fact, just about the design luminosities being talked about for ISABELLE.

Much more spectacular than the reaction (11.1) is the direct formation of the Z boson,

$$e^+ + e^- \to Z \to \text{anything in the world} , \tag{11.3}$$

at electron-positron colliding-beam machines of the kind being discussed at CERN with $\sqrt{s}$ up to 140 GeV, tentatively named LEP. In this reaction the Z boson is formed at rest and subsequently decays into all kinds of lepton pairs and quark pairs, which, in turn, materialize into hadron jets. The final states here are nothing other than the modern version of Earth, Water, Fire, and Air. The production cross section exactly at the Z peak can be easily estimated. In terms of R, the cross section in units of the one-photon-exchange muon-pair cross section, we obtain, in the three-lepton doublet, six-quark model,

$$R_{\text{visible}} = (12\pi/m_Z{}^2) B_{e^+e^-} B_{\text{visible}} / (4\pi\alpha^2/3m_Z{}^2) \simeq 5000 \tag{11.4}$$

where R_{visible} means that we exclude the $\nu\bar{\nu}$ channels. Note that the famous R here is not 5, not 50, but 5000. The total decay width of the Z boson is predicted to be about 1.5 GeV for $m_Z \simeq 80$ GeV.

It is fun to speculate what may happen in about five to ten years. We consider here two extreme options:

Scenario 1: Exhaustive searches for weak bosons have revealed no $W^\pm$, nor Z^0. From a certain theoretical point of view a world without W and Z is at least as interesting as the "orthodox" world. The nonexistence of W and Z means the absence of simple poles in the current channels but arguments based on dimensions assure us that $q\bar{q}$ and $\ell\bar{\ell}$ scattering are predicted to be strong in any case at $\sqrt{s} \simeq 100$ - 300 GeV.[62] Presumably there will be complicated cuts in the current channels. Could the whole weak interaction theory be like the S matrix theory à la Chew of the '60s? Perhaps nature may be aware of tricks even 't Hooft doesn't know.

Scenario 2: The W and Z bosons are found exactly at the masses predicted by Weinberg. We may naturally ask: Does this prove SBGT? The heretic opens his notebook of 1978 and recalls that the Weinberg mass relations are possible even in the alternative model based on γW^0 mixing (see the "unification point" of Fig. 10) provided we require asymptotic $SU(2) \otimes U(1)$, i.e., we demand that $SU(2) \otimes U(1)$ be unbroken at high q^2. To be more precise, let us try to arrange λ, e, and g of the alternative model in such a way that at high q^2 no $SU(2) \otimes U(1)$ violating terms like $J_\lambda^3 J_\lambda^Y$ are present in the effective Lagrangian. This can be easily done; we obtain, for $q^2 >> m_Z^2$,

$$\frac{1}{2} J^\dagger \Delta J \simeq \frac{1}{2} \frac{1}{q^2} \left[\frac{e^2}{1-\lambda^2} J_\lambda^Y J_\lambda^Y + g^2 J_\lambda^3 J_\lambda^3 \right] \tag{11.5}$$

provided the mixing parameter λ satisfies

$$\lambda = e/g \quad . \tag{11.6}$$

When we insert this condition in the mass relation (10.5b) and use (10.8) and (10.11), we immediately obtain the Weinberg mass relations (4.2). It can also be shown that the "unification condition" (11.6) insures decent high-energy behavior in $\nu + \bar{\nu} \to W^+ + W^-$ in the sense of tree unitarity and makes the g factor of $W^\pm$ boson magnetic moment agree with the Glashow[17]-Salam-Ward[18]-Weinberg[31] value,[101] 2.

From the heretic's point of view the Weinberg mass relations, if confirmed, would appear to be a major triumph of $SU(2) \otimes U(1)$ symmetry, and not so much of SBGT. The relevance of asymptotic $SU(2) \otimes U(1)$ to the Weinberg mass relations was first emphasized by Bjorken.[62]

XII. HIGGS HUNTING

By this time the true believer of the orthodoxy has become rather impatient. He asks: What is really needed to prove SBGT? The answer is simple: we must check whether the symmetry is indeed "spontaneously broken" in the manner required by the Higgs mystique.[26,27]

SBGT needs the Higgs boson. An integral part of the orthodox dogma is that not only m_W but also m_F, where F stands for any fermion, lepton, or quark, is zero before we let $\phi \to \phi - \langle\phi\rangle_{vac}$. The nonvanishing vacuum expectation value of the Higgs field is responsible for not just the gauge boson masses but also the fermion masses. This means that the Higgs couplings to various particles are directly proportional to their masses:

$$\begin{aligned} m_W^2 &= \frac{1}{4} g^2 \, |\langle\phi\rangle_{vac}|^2 \quad , \\ m_F^2 &= g_{HFF}^2 \, |\langle\phi\rangle_{vac}|^2 \quad . \end{aligned} \tag{12.1}$$

But g^2 is related to Fermi's G via (10.8); so, eliminating $\langle\phi\rangle_{vac}$, we have

$$g_{HFF}^2 = \sqrt{2}\, G\, m_F^2 \quad , \tag{12.2}$$

at least in the simplest SBGT model with just one Higgs doublet. This is a remarkable result. We don't know whether the Higgs boson is as light as 5 GeV[102] or as heavy as 200 GeV but we do know g_{HFF}^2 for the Higgs coupling to any lepton and quark very precisely. Why is the muon 207 times heavier than the electron? Answer: the Higgs boson is coupled to the muon 207 times more strongly--a simple question, a simple answer! Note that we expect a violent violation of μe universality, $\tau\mu$ universality, etc.

Several people have discussed how we might look for the Higgs boson.[103] If its mass is below 9 GeV, it may be looked for in Υ (9.4) decay:

$$\Upsilon \to H + \gamma \tag{12.3}$$

as suggested by Wilczek.[104] See Fig. 11(a). This is a favorable

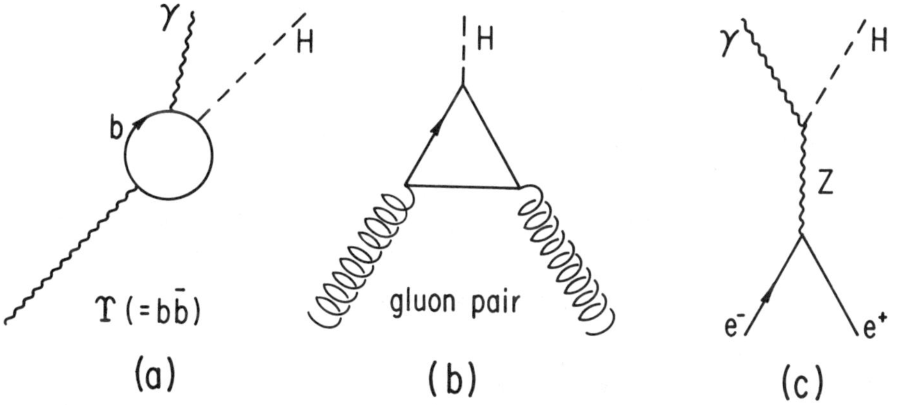

Fig. 11. Production of the Higgs boson.

situation because the Higgs boson is supposed to be coupled more strongly to the heavy quarks. Assuming that Υ is a $(b\bar{b})$ bound state with $m_b \simeq 4.5$ GeV, for which we have no conclusive evidence as of this conference, we expect

$$\frac{\Gamma(\Upsilon \to H + \gamma)}{\Gamma(\Upsilon \to \mu^+ + \mu^-)} = 0.007 \times \text{phase space} \quad . \tag{12.4}$$

This decay mode may be studied at DORIS, CESR, and PETRA, which are expected to be excellent Υ factories, by detecting mono-energetic γ rays in

$$e^+ + e^- \to \Upsilon \to \gamma + H \quad . \tag{12.5}$$

Another possibility is to study

$$p + p \rightarrow H + \text{any} \qquad (12.6)$$

where the Higgs boson is assumed to be formed in "gluon-gluon collisions" [see Fig. 11(b)]. Cross section estimates[105] show that such an experiment is probably within the realm of feasibility at colliding-beam facilities like ISABELLE. There are other reactions mentioned in the literature:

$$\begin{aligned} e^+ + e^- &\rightarrow Z + H \quad , \\ e^+ + e^- &\rightarrow Z \rightarrow H + \mu^+ + \mu^- \quad . \end{aligned} \qquad (12.7)$$

Once produced the Higgs boson will have very characteristic decay signatures. Because g^2_{HFF} is proportional to $m_F{}^2$, it will decay preferentially into heavier objects like $\tau^+ + \tau^-_+$, new heavy hadrons, etc. If sufficiently massive, it decays into $W^+ + W^-$ and $2Z^0$, producing utterly spectacular jets.

It has been pointed out by the late Ben Lee and others[106] that, if the Higgs boson turns out to be very heavy, our theoretical views on high-energy weak interactions must be modified. Their argument, in its simplest form, goes as follows. The Higgs boson is massive; so it is coupled also to itself. The strength of the HHH self-coupling is enormous if the Higgs boson is very heavy. The reason is that its dimensionless coupling constant is of the order of $Gm_H{}^2$, which becomes a "strong interaction constant" for

$$m_H \gtrsim 1/\sqrt{G} \simeq 300 \text{ GeV} \quad . \qquad (12.8)$$

With such a large constant, perturbation theory must obviously break down, and we even expect "Phenomenological Higgs" as bound states.

For successes of the renormalization programs, tree unitarity, etc., a strong HHH coupling does not seem to be desirable. If SBGT is to be of any use, the perturbation expansions must converge rapidly and higher-order diagrams must give smaller contributions just as in QED. This means, in particular, that weak scattering processes--$q\bar{q}$, $\ell\bar{\ell}$, WW, and HH scattering--must all remain weak except in the vicinity of W, Z, and H "resonances." It is of utmost importance to test this point at future colliding-beam facilities. We may conceive of other detailed tests of SBGT like checking complicated loop diagrams--are the triangular anomalies really cancelled? --but such tests may be hopeless in practice short of precision experiments comparable to the g-2 experiment for the muon.

XIII. FUTURE OUTLOOK

What can we look forward to before we enter the 21st Century?

We can certainly expect progress in quark, lepton spectroscopy; more quark flavors and possibly more heavy leptons. Will the proliferation of quarks terminate as required by QCD satisfying asymptotic freedom?[107] Or we just continue with "one, two, three,... infinity"? Will we ever understand the quark, lepton spectrum?

Coming more closely to the main theme of this talk, we may ask: When the new quarks and leptons participate in the weak interactions, do they always form left-handed doublets and right-handed singlets, never right-handed doublets? What are the carriers of weak forces? Are they $W^{\pm}$ and Z^0, or something unexpectedly different? Do the W and Z interact in such a way that the weak and electromagnetic forces get unified in the sense of $SU(2) \otimes U(1)$? Will the orthodox view à la Salam, Weinberg,... be vindicated, or will nature reveal surprises--like low-mass W and Z, or even no W, Z, H at all?

The physics of the 20th Century opened with the emergence of the photon as the carrier of electromagnetic forces; this line of development, pioneered by Planck and Einstein, culminated in the 1927 paper entitled "The Quantum Theory of the Emission and Absorption of Radiation" of Professor Dirac.[5] It is fitting and proper that in the last quarter of this century we are engaged in the colossal task of identifying the basic carriers of weak forces, $W^{\pm}$, Z^0 (or something else). If there are no world wars, no world-wide economic depression, no take-over of the civilized world by terrorists, etc., super-high-energy machines like LEP (e^+e^- up to $\sqrt{s}$ = 140 GeV) will be constructed before long, and by the year 2000 we will have the answers to the burning questions of the '70s.

REFERENCES

1. P. A. M. Dirac, Proc. Roy. Soc. London, A 117, 610 (1928).
2. Pauli's suggestion appeared in an open letter dated December 4, 1930, and addressed to "Lieber Radioaktive Damen und Herren" of Tübingen. The letter is reproduced in Collected Scientific Papers by Wolfgang Pauli (ed. R. Kronig and V. F. Weisskopf; Interscience, New York, 1964), Vol. 2, p. 1313.
3. E. Fermi, Ric. Scientifica 4, 491 (1933); Nuovo Cimento 2, 1 (1934).
4. E. Fermi, Z. Physik 88, 161 (1934).
5. P. A. M. Dirac, Proc. Roy. Soc. London A 114, 243 (1927).
6. P. Jordan and E. Wigner, Z. Physik 47, 631 (1928).
7. H. Yukawa, Proc. Phys.-Math. Soc. Japan 17, 48 (1935).
8. R. P. Feynman and M. Gell-Mann, Phys. Rev. 109, 193 (1958); E. C. G. Sudarshan and R. E. Marshak, Phys. Rev. 109, 1860 (1958); J. J. Sakurai, Nuovo Cimento 7, 649 (1958).
9. N. Cabibbo, Phys. Rev. Lett. 10, 531 (1963).
10. M. Gell-Mann and M. Lévy, Nuovo Cimento 16, 705 (1960).
11. C. L. Cowan et al., Science 124, 103 (1956).
12. G. Danby et al., Phys. Rev. Lett. 9, 36 (1962).
13. J. Schwinger, Ann. Phys. 2, 407 (1957).
14. S. A. Bludman, Nuovo Cimento 9, 443 (1958).
15. C. N. Yang and R. L. Mills, Phys. Rev. 96, 191 (1954).
16. M. Gell-Mann, Proceedings of the 1960 Annual International Conference on High Energy Physics at Rochester (ed. E. C. G. Sudarshan, J. H. Tinlot, and A. C. Melissinos; University of Rochester, 1960), p. 508.

17. S. L. Glashow, Nucl. Phys. 22, 579 (1961).
18. A. Salam and J. C. Ward, Phys. Lett. 13, 168 (1964).
19. S. B. Treiman, Nuovo Cimento 15, 916 (1960); T. D. Lee and C. N. Yang, Phys. Rev. 119, 1410 (1960).
20. T. D. Lee and C. N. Yang, Phys. Rev. Letters 4, 307 (1960).
21. N. Cabibbo and R. Gatto, Phys. Rev. 124, 1577 (1961).
22. D. C. Cundy et al., Phys. Lett. 31B, 423 (1972).
23. A. F. Rothenberg, SLAC-147 (1972).
24. S. Weinberg, Phys. Rev. Lett. 19, 1264 (1967).
25. A. Salam, Elementary Particle Theory (ed. N. Svartholm; Almquist and Wiksells, Stockholm, 1968)p. 367.
26. P. W. Higgs, Phys. Lett. 12, 132 (1964). Phys. Rev. Lett. 13, 508 (1964).
27. F. Englert and R. Brout, Phys. Rev. Lett. 13, 321 (1964); G. S. Guralnik, C. R. Hagen, and T. W. B. Kibble, Phys. Rev. Lett. 13, 585 (1964).
28. The idea of generating gauge boson masses by introducing neutral scalar fields goes back to the 1957 paper of Schwinger (Reference 13).
29. G. 't Hooft, Nucl. Phys. B33, 173 (1971); B35, 167 (1971).
30. For the renormalizability of SBGT see also B. W. Lee and J. Zinn Justin, Phys. Rev. D5, 3121, 3137, 3155 (1972); G. 't Hooft and M. Veltman, Nucl. Phys. B44, 189 (1972).
For the quantization of the Yang-Mills theory see, for example, R. P. Feynman, Acta Phys. Polon. 26, 697 (1963); B. deWitt, Phys. Rev. 162, 1195, 1239 (1967); L. D. Faddeev and V. N. Popov, Phys. Lett. 25B, 29 (1967).
The history of this subject is reviewed in M. Veltman, Proceedings of the VI International Symposium on Electron and Photon Interactions at High Energies, Bonn, August 1973 (ed. H. Rollnik and W. Pfeil; North Holland, Amsterdam, 1974), p. 429.
31. S. Weinberg, Phys. Rev. Lett. 27, 1688 (1971); J. S. Bell, Nucl. Phys. B 60, 427 (1973); C. H. Llewellyn Smith, Phys. Lett. 46B, 233 (1973); J. M. Cornwall, D. Levin, and G. Tiktopoulos, Phys. Rev. Lett. 30, 1268 (1973).
32. M. Gell-Mann et al., Phys. Rev. 179, 1518 (1969).
33. See, e.g., Particle Data Group, Revs. Mod. Phys. 48, S1 (1976).
34. S. L. Glashow, J. Iliopoulos, and L. Maiani, Phys. Rev. D2, 1285 (1970).
35. For earlier discussions of a fourth quark see e.g., Y. Hara, Phys. Rev. 134, B701 (1964); J. D. Bjorken and S. L. Glashow, Phys. Lett. 11, 255 (1964); Z. Maki and Y. Ohnuki, Progre. Theor. Phys. 32, 144 (1964).
36. For review see e.g., B. Richter, Rev. Mod. Phys. 49, 251 (1977); S. C. C. Ting, Rev. Mod. Phys. 49, 235 (1977).
37. J. Ellis, CERN preprint, TH 2365 (1977).
38. F. Hasert et al., Phys. Lett. 46B, 138 (1973); Nucl. Phys. B73, 1 (1974).
39. G. Mayatt, Proceedings of the VI International Symposium on Electron and Photon Interactions at High Energies, Bonn, August 1973 (ed. H. Rollnik and W. Pfeil; North Holland, Amsterdam, 1974), p. 389.

40. A. Benvenuti et al., Phys. Rev. Lett. 32, 800 (1974).
41. B. C. Barish et al., Phys. Rev. Lett. 34, 538 (1975).
42. S. J. Barish et al., Phys. Rev. Lett. 33, 448 (1974).
43. F. Hasert et al., Phys. Lett. 46B, 121 (1973).
44. The formulas for $R_{\nu N}$, $R_{\bar{\nu} N}$ in terms of $\sin^2\theta_W$ were derived by many authors: L. M. Sehgal, Nucl. Phys. B 65, 141 (1973); Phys. Lett. 48B, 132 (1974); R. B. Palmer, Phys. Lett. B 46, 240 (1973); A. De Rújula et al., Rev. Mod. Phys. 46, 391 (1974). See also A. Pais and S. B. Treiman, Phys. Rev. D6, 2700 (1972); E. A. Paschos and L. Wolfenstein, Phys. Rev. D7, 91 (1973).
45. P. Q. Hung and J. J. Sakurai, Phys. Lett. 63B, 295 (1976); 69B, 323 (1977); 72B, 208 (1977).
46. G. Rajasekaran and K. V. L. Sarma, Pramana 2, 62 (1974); G. Ecker and H. Pietschmann, Acta Phys. Austr. 45, 313 (1976); G. Ecker, Phys. Lett. 72B, 450 (1978); J. D. Bjorken, *Proceedings of the Summer Institute on Particle Physics*, SLAC, August 1976 (ed. M. C. Zipf; SLAC Stanford, 1976), p. 1.
47. The importance of model-independent analyses has also been emphasized by A. Pais and S. B. Treiman, Phys. Rev. D9, 1459 (1974); L. Wolfenstein, AIP Proceedings 23, 84 (1974).
48. In principle the diffractive production of ϕ and ψ/J by weak neutral currents could determine the vector coupling constants of $\bar{s}s$ and $\bar{c}c$.
49. The possibility that the neutral-current interactions involve SPT will not be entertained here; see e.g., B. Kayser et al., Phys. Lett. 52B, 385 (1974); R. L. Kingsley, F. Wilczek, and A. Zee, Phys. Rev. D 10, 2216 (1974).
50. We expect, however, small corrections to (6.42) due to $\bar{s}s$, $\bar{c}c$ ... pairs.
51. For Gargamelle see J. Blietschau et al., Nucl. Phys. B118, 218 (1977). For HPWF (Harvard-Pennsylvania-Wisconsin-Fermilab) see P. Wanderer et al., preprint HPWF-77/1 (1977). For Caltech see F. S. Merritt et al., Caltech preprint CALT 68-601 (1977). For CHDS (CERN-Heidelberg-Dortmund-Saclay) see M. Holder et al., Phys. Lett. 71B, 415 (1977).
52. R. P. Feynman, *Photon Hadron Interactions* (W. A. Benjamin, New York, 1972), p. 132; S. M. Berman, J. D. Bjorken, and J. B. Kogut, Phys. Rev. D4, 3338 (1971).
53. C. H. Albright and J. Cleymans, Nucl. Phys. B76, 48 (1974); L. M. Sehagal, Nucl. Phys. B90, 471 (1974); J. Okada and S. Pakvasa, Nucl. Phys. B112, 400 (1976); P. Q. Hung, Phys. Lett. 69B, 216 (1977).
54. R. D. Field and R. P. Feynman, Phys. Rev. D15, 2590 (1977). For an excellent review on applications of the quark fragmentation hypothesis to lepton-induced reactions, see L. M. Sehgal, Aachen preprint, PITHA-81 (1977).
55. H. Kluttig, J. A. Morfín, and W. Van Doninck, Phys. Lett. 71B, 446 (1977).
56. The presence of isoscalar-isovector interference was established also in $\nu p(n) \rightarrow \nu p(n)\pi^+\pi^-$; the effect observed is about 2-1/2 standard deviations away from no isoscalar-isovector

interference. N. P. Samios (private communication) based on work of the BNL 7 ft Bubble Chamber Group.
57. L. M. Sehgal, Phys. Lett. 71B, 99 (1977).
58. P. Langacker and D. P. Sidhu, University of Pennsylvania preprint, UPR-0085T (1978).
59. D. Cline et al., Phys. Rev. Lett. 37, 252 (1976); 37, 648 (1976).
60. W. Krenz et al., CERN/EP-PHYS-77-52 (1977); O. Erriques et al., Phys. Lett. 73B, 350 (1978). These groups also present further evidence for isoscalar-isovector interference.
61. Such a model was proposed by M. A. B. Bég and A. Zee, Phys. Rev. Lett. 30, 675 (1973); V. S. Mathur, S. Okubo, and J. E. Kim, Phys. Rev. D11, 1059 (1975).
62. J. D. Bjorken, SLAC-PUB-2062 (1977).
63. See e.g., H. Faissner et al., Aachen preprint PITHA-91 (1977); M. Gourdin, Proceedings of the International Neutrino Conference, Aachen 1976 (ed. H. Faissner, H. Reithler, and P. Zerwas; Vieweg, Braunschweig, 1977), p. 234.
64. Such a plot was first made by H. H. Chen and B. W. Lee, Phys. Rev. D5, 1874 (1972).
65. F. Reines, H. S. Gurr, and H. W. Sobel, Phys. Rev. Lett. 37, 315 (1976).
66. J. Blietschau et al., Nucl. Phys. B 114, 189 (1976).
67. This may be somewhat disturbing because the preferred value of $\sin^2\theta_W$ from the inelastic data of the CHDS Collaboration (Reference 51) is around 0.25.
68. M. A. Bouchiat and C. Bouchiat, Phys. Lett. 48B, 111 (1974); I. B. Khriplovich, JETP Lett. 20, 315 (1974). For gauge model implications see J. Bernabéu and C. Jarlskog, Nuovo Cimento 38A, 295 (1977).
69. L. L. Lewis et al., Phys. Rev. Lett. 39, 795 (1977).
70. P. E. Baird et al., Phys. Rev. Lett. 39, 798 (1977).
71. E. M. Henley and L. Wilets, Phys. Rev. A14, 1411 (1976); I. B. Khriplovich, V. N. Novikov, and O. P. Suchkov, Sov. Phys. JETP 44, 872 (1976).
72. L. M. Barkov and M. S. Zolotoryov, Pisma ZHETF 27, 379 (1978) [JETP Letters (to be published)].
73. S. L. Adler and S. F. Tuan, Phys. Rev. D11, 129 (1975).
74. H. Fritzsch, M. Gell-Mann, and P. Minkowski, Phys. Lett. 59B, 256 (1975); S. Pakvasa, W. A. Simmons, and S. F. Tuan, Phys. Rev. Lett. 35, 702 (1975.
75. H. Faissner et al., Proceedings of the International Neutrino Conference, Aachen 1976 (ed. H. Faissner, H. Reithler, and P. Zerwas; Vieweg, Braunschweig, 1977), p. 278.
76. P. Fayet, Nucl. Phys. B 78, 14 (1974); P. Ramond, Nucl. Phys. B 110, 214 (1976); Y. Achiman, K. Koller, and T. F. Walsh, Phys. Lett. 59B, 261 (1975); R. M. Barnett, Phys. Rev. Lett. 34, 41 (1975); F. Gürsey and P. Sikivie, Phys. Rev. Lett. 36, 775 (1976).
77. B. W. Lee and S. Weinberg, Phys. Rev. Lett. 38, 1237 (1977).
78. J. Bernabéu and C. Jarlskog, Phys. Lett. 69B, 71 (1977).
79. T. P. Cheng and L. F. Li, Phys. Rev. Lett. 38, 381 (1977).

80. P. Langacker and G. Segrè, Phys. Rev. Lett. 39, 259 (1977).
81. H. Fritzsch and P. Minkowski, Nucl. Phys. B103, 61 (1976); R. Mohapatra and D. P. Sidhu, Phys. Rev. Lett. 38, 667 (1977).
82. Yu. V. Gapanov and I. V. Tyutin, Soviet Phys. JETP 20, 1231 (1965); A. Ali and C. A. Dominguez, Phys. Rev. D12, 3673 (1975).
83. H. S. Gurr, F. Reines, and H. W. Sobel, Phys. Rev. Lett. 33, 179 (1974).
84. T. W. Donnelly et al., Phys. Lett. 49B, 8 (1974); G. J. Gounaris and J. D. Vergados, Phys. Lett. 71B, 35 (1975).
85. A. Le Yaouanc et al., Nucl. Phys. B125, 243 (1977).
86. J. E. Kim, P. Langacker, and S. Sarkar, University of Pennsylvania preprint, UPR 0086 (1978).
87. L. F. Abbott and R. M. Barnett, SLAC-PUB-2097 (1978).
88. E. H. Monsay, ANL-HEP-PR-78-08 (1978).
89. S. L. Adler, Ann. Phys. 50, 189 (1968); Phys. Rev. D9, 229 (1974); Phys. Rev. D12, 2644 (1975).
90. The ratios quoted here are from Monsay's paper (Reference 88); Abbott and Barnett also obtain similar results (private communication).
91. F. J. Hasert et al., Phys. Lett. 59B, 485 (1975).
92. W. Lee et al., Phys. Rev. Lett. 38, 202 (1977).
93. G. Feinberg and M. Y. Chen, Phys. Rev. D10, 190 (1974). R. R. Lewis and W. L. Williams, Phys. Lett. 59B, 70 (1975); R. N. Cahn and G. L. Kane, Phys. Lett. 71B, 348 (1977).
94. A. Love, G. G. Ross, and D. V. Nanopoulos, Nucl. Phys. B49, 513 (1972); E. Derman, Phys. Rev. D7, 2755 (1973); M. Suzuki, Nucl. Phys. B70, 154 (1974); S. M. Berman and J. R. Primack, Phys. Rev. D9, 217 (1974); M. A. B. Bég and G. Feinberg, Phys. Rev. Lett. 33, 606 (1974); R. N. Cahn and F. J. Gilman, Phys. Rev. D17, 1313 (1978).
95. T. Kinoshita et al., Phys. Rev. D2, 910 (1970); J. Godine and A. Hankey, Phys. Rev. D6, 3301 (1972); R. Bundy, Phys. Lett. 45B, 340 (1973).
96. P. Q. Hung and J. J. Sakurai, UCLA preprint UCLA/78/TEP/8 (1978).
97. L. Yu. Kobzarev, L. B. Okun, and I. Ya. Pomeranchuk, Soviet Physics-JETP 14, 355 (1962); G. Feldman and P. T. Matthews, Phys. Rev. 132, 823 (1963); S. Coleman and H. J. Schnitzer, Phys. Rev. 134, B863 (1964).
98. S. D. Drell and T.-M. Yan, Phys. Rev. Lett. 25, 316 (1970).
99. Y. Yamaguchi, Nuovo Cimento 43, 193 (1966); F. Chilton, A. M. Saperstein, and E. Shrauner, Phys. Rev. 148, 1380 (1966).
100. C. Quigg, Rev. Mod. Phys. 49, 317 (1977); L. M. Lederman and B. G. Pope, Phys. Rev. Lett. 27, 765 (1971); R. B. Palmer et al., Phys. Rev. D14, 118 (1976).
101. For pedagogical discussion on the W boson g factor see e.g., J. Bernstein, Rev. Mod. Phys. 46, 7 (1974).
102. Some theorists argue that the Higgs boson cannot be too light: A. D. Linde, JETP Letters 23, 73 (1976); S. Weinberg, Phys. Rev. Lett. 36, 294 (1976); P. H. Frampton, Phys. Rev. Lett. 37B, 1378 (1976).
103. For earlier discussion see e.g., J. Ellis, M. K. Gaillard, and D. V. Nanopoulos, Nucl. Phys. B106, 292 (1976).

104. F. Wilczek, Phys. Rev. Lett. 39, 1304 (1977).
105. H. M. Georgi et al., Harvard preprint HUTP-77/A084 (1977).
106. B. W. Lee, C. Quigg, and H. B. Thacker, Phys. Rev. Lett. 38, 883 (1977); M. Veltman, Phys. Lett. 70B, 253 (1977).
107. H. D. Politzer, Phys. Rev. Lett. 30, 1346 (1973); D. J. Gross and F. Wilczek, Phys. Rev. Lett. 30, 1343 (1973).

APPLICATIONS OF QCD*

John Ellis
CERN - Geneva, Switzerland
and
Stanford Linear Accelerator Center
Stanford University, Stanford, California 94305

Talk presented at the meeting on "Current Trends in the Theory of Fields" held in honor of Prof. P.A.M. Dirac on the occasion of his 75th birthday and the 50th anniversary of the Dirac equation, at the Florida State University, Tallahassee, Florida 32306, April 6th and 7th, 1978.

---

*Work supported by the Department of Energy

ISSN: 0094-243X/78/081/$1.50

1. INTRODUCTION

A substantial fraction of the theoretical physics community tends to be rather smug these days. It feels that not only are the weak and electromagnetic interactions known to be combined in a non-abelian gauge theory, but that the theory of the strong interactions is known to be Quantum Chromodynamics (QCD).[1] It is true that there are one or two minor technical problems to be clarified, such as the mechanism by which quarks are confined (if indeed they are), but the riddles of the nuclear interactions are supposed to be solved in principle. All very fine, but the rest of the community is entitled to ask for the positive evidence that QCD is correct. At present much of the evidence is either by default (no other plausible field theory seems capable of asymptotic freedom[2] or confinement) or purely aesthetic (QCD is beautiful, it is a gauge theory as are our theories of the other fundamental interactions) or rather qualitative (charmonium,[3] the approximate validity of the parton model,[4] the Zweig rule,[5] and so on). The community is surely entitled to see some direct experimental confirmation of specific theoretical numbers reliably predicted by QCD.

The purpose of this talk is to review the status of reliable QCD predictions which either have, or soon can be verified by experiment. It is divided into three parts:

1. A discussion of the classic application of QCD perturbation theory and asymptotic freedom to predict scaling violations in deep inelastic leptoproduction experiments.[6] Emphasis will be laid on recent results from the BEBC neutrino collaboration[7] which provide the first direct experimental confirmation of the numerical values of the anomalous dimensions predicted[6] by QCD. These data may constitute the best phenomenological evidence to date in favor of QCD.

2. A review of recent advances in developing and justifying QCD perturbation theory predictions for a number of physical applications not underwritten by the operator product expansion and renormalization group arguments generally used to motivate the application of asymptotic freedom to deep inelastic processes. These include predictions for two- and multi-jet production cross-sections in e^+e^-, lepton-hadron and hadron-hadron collisions[8,9]; a modified Drell-Yan formula for lepton-pair production in hadron-hadron collisions[10]; and scaling violations in hadronic final states[11] analogous to those seen in deep inelastic structure functions. Very few of these predictions have yet been confronted with experiment, but they promise to provide copious, precise and reliable ways to verify or disprove QCD.

3. A final mention will be made of attempts to address the question whether these predictions of QCD perturbation theory should be regarded as reliable, given the fact that non-perturbative effects[12] are presumably crucial in QCD (to confine quarks,[13] for example). Analyses[14] of the simplest non-perturbative corrections to the most basic deep inelastic process $\sigma(e^+e^- \to \gamma^* \to \text{hadrons})$ indicate that they are negligible at large momentum transfers. This suggests that QCD perturbation theory predictions may indeed by reliable for the deep inelastic, large momentum transfer processes where the previously proposed experimental tests are to be made.[15]

2. DEEP INELASTIC SCATTERING

The classic tests of QCD are afforded by deep inelastic leptoproduction. Every field theory predicts that the structure functions should violate scaling as $Q^2\to\infty$, but QCD makes very specific predictions, owing to the fact that its coupling constant is asymptotically free at large momenta[16]:

$$\alpha_s(Q^2) \equiv \frac{g^2(Q^2)}{4\pi} \underset{Q^2\to\infty}{\simeq} \frac{12\pi}{(33-2N_f)\ln(Q^2/\Lambda^2)} \tag{1}$$

These predictions are expressible most precisely in terms of the Q^2 dependence expected for the moments[17]

$$M_i(N,Q^2) \equiv \int_0^1 dx\, x^{N-2}\, F_i\,(x,Q^2) \tag{2}$$

of the deep inelastic structure functions F_i (F_2 and xF_3 will be discussed here). QCD predicts[6] that the moments $M_i(N,Q^2)$ should behave as negative powers of $\log(Q^2/\Lambda^2)$ as $Q^2\to\infty$, whereas any other field theory would predict power-law violations of scaling.[18] The predictions are simplest for flavor non-singlet structure functions, such as $F_2^{ep}-F_2^{en}$, or the vector-axial vector interference structure function xF_3 measured in charged current νN and $\bar{\nu}N$ scattering. For these cases

$$M_i(N,Q^2) \underset{Q^2\to\infty}{\simeq} C_i^N(\ln Q^2/\Lambda^2)^{-d_N}\left[1+O\left(\frac{1}{\log Q^2/\Lambda^2},\frac{\log\log Q^2/\Lambda^2}{\log Q^2/\Lambda^2}\right)\right] \tag{3}$$

where the powers d_N were calculated[6] to be

$$d_N = \frac{4}{33-2N_f}\left[1-\frac{2}{N(N+1)}+4\sum_{j=2}^{N}\frac{1}{j}\right] \tag{4}$$

The predictions for flavor singlet structure functions are more complicated,[6] with two leading terms differing by less than one power of $\log Q^2/\Lambda^2$ in their asymptotic behavior as $Q^2\to\infty$.

The results (1) to (4) are guaranteed[1,6,19] by the renormalization group to be the result of summing all the logarithms encountered in QCD perturbation theory. The moments (2) have the effect of picking out the matrix elements of operators of definite spin N[17] in the operator product expansion of two electromagnetic or weak moments. The power d_N is the anomalous dimension of scaling of the spin N operator. It results from exponentiation of the simple vertex corrections to the quark-antiquark operator indicated in Fig. 1a. The behaviors of the moments of singlet structure functions are more complicated because there are two operators of the same spin N, a two-gluon operator as well as a quark-antiquark operator, and they mix together through diagrams like that in Fig. 1b.

While the predictions (2) to (4) are precise, it is often convenient to reexpress them in terms of the evolution with Q^2 of the distribution of effective quark (or gluon) partons within the hadron target.[20] This can be done by inverting the expressions (3) and (4) for the moments $M_i(N,Q^2)$ by using a Mellin Transform.[21] The physics of the ensuing distributions $q(x,Q^2)$ (or $g(x,Q^2)$) can be seen very clearly from Fig. 1 when we recall that x is[4] the longitudinal momentum fraction carried by the parton. The anomalous dimensions d_N are determined by the basic vertices of Fig. 2, which cause the parton distributions to evolve either by bremsstrahlung (Fig. 2a) or by pair creation (Fig. 2b). The equation controlling the evolution of $q(x,Q^2)$ at large Q^2 is just[22]

$$Q^2\frac{dq_i}{dQ^2}(x,Q^2) = \frac{\alpha_s(Q^2)}{2\pi}\int_x^1 dy \left[q_j(y,Q^2)P_{q_j\to q_i}\left(\frac{x}{y}\right) + g(y,Q^2)P_{g\to q_i}\left(\frac{x}{y}\right)\right] \quad (5)$$

and the evolution of $g(x,Q^2)$ obeys an analogous equation.[22] The "splitting" functions $P_{q\to q}$, $P_{g\to q}$, $P_{g\to g}$, $P_{q\to g}$ are easily calculable from the vertices of Fig. 2, and analogous to (for example) the equivalent photon distribution in Weiszäcker-Williams calculations in good old QED. As an example,[22]

$$P_{q_j\to q_i}(z) = \frac{4}{3}\,\delta_{ij}\left[\frac{1+z^2}{(1-z)_+} + \frac{3}{2}\,\delta(z-1)\right] \quad (6)$$

and so on. The δ function in equation (6) just normalizes the fermion number to 1. The anomalous dimensions d_N can easily be reconstructed by taking the moments of equation (5), and are given by the z moments of the "splitting" functions like $P_{q\to q}$ in equation (6).

There are two reasons for focusing on the Q^2-dependent effective proton distributions $q(x,Q^2)$ and $g(x,Q^2)$. One reason is that they are the quantities directly related to the measured deep inelastic structure functions, and hence very convenient objects to work with phenomenologically. The other reason is that one might hope that they would have more universal applicability, and we will indeed see in the next section that recent theoretical calculations indicate that the same effective parton distributions can be used in many other phenomenological applications, such as in calculating hard quark-quark scattering contributions to large p_T hadron production, for example.

One can anticipate directly from the diagrams of Fig. 2 certain qualitative features[20] of the development of $q(x,Q^2)$ with increasing Q^2. Clearly both the bremsstrahlung and pair creation processes tend to degrade the parton longitudinal momenta as Q^2 is increased, and this can be seen directly by substituting the "splitting" function $P_{q\to q}(z)$ (6) into the evolution equation (5). Therefore deep inelastic structure functions will tend to fall in towards x=0 as $Q^2\to\infty$, in a manner indicated qualitatively in Fig. 3.[20] This behavior can also be seen in the form of the anomalous dimensions d_N (4). As $N\to\infty$, $d_N \propto \log N$, so that higher moments $M_i(N,Q^2)\to 0$ faster as $N\to\infty$. But the higher moments are seen from equation (2) to weight larger values of x closer to 1, and we see once more that the structure function at large x should fall to zero as $Q^2\to\infty$.

This trend is indeed seen in the data[23-25]: Figure 4 shows recent data from a FNAL muon scattering experiment, and superimposing data from different ranges of Q^2 clearly manifests the qualitative behavior that we anticipated in Fig. 3. On the other hand, almost any field theory would predict an analogous fall in towards x=0,[20]

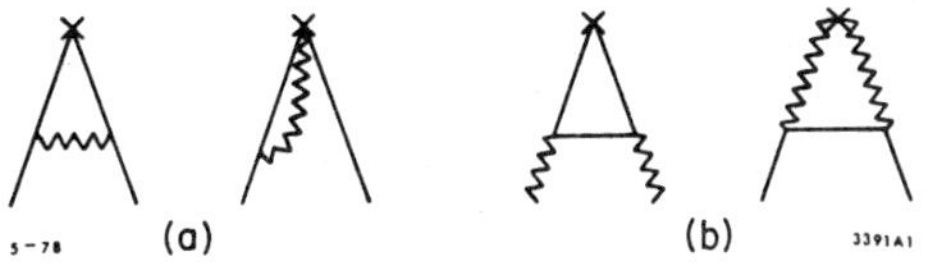

Fig. 1. Some diagrams contributing (a) to the anomalous dimensions of quark-antiquark operators, (b) to the mixing of flavor singlet operators.

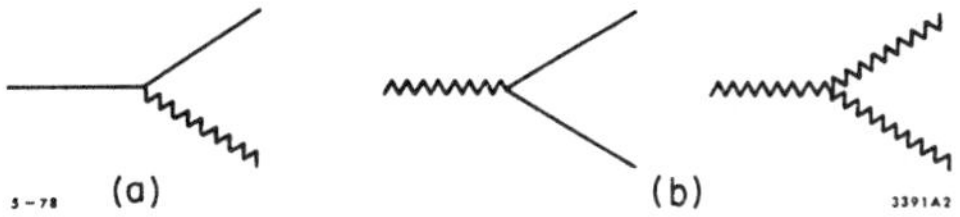

Fig. 2. The basic vertices for (a) gluon bremsstrahlung and (b) for q-$\bar{q}$ and gluon pair creation, which control the anomalous dimensions and the evolution of the quark and gluon distributions.

because basic vertices analogous to the bremsstrahlung and pair creation creation graphs of Fig. 2 exist in almost any field theory. What is specific to QCD is the characteristic logarithmic Q^2 dependence characterized by equations (1) to (4).

To compare these predictions with experimental data, several groups[26,27] have constructed parametrizations of effective quark and gluon distributions which are consistent with the QCD moment equations (2) to (4) (or equivalently the evolution equations (5) and (6)) and made phenomenological fits to the experimental data on eN, μN and νN scattering. Sample graphs from one such analysis[27] are displayed in Fig. 5: the fits seem to work at least semi-quantitatively. The particular graphs shown have been selected because they are the only ones made with a complete analysis[28] of the $O(^1/\log Q^2/\Lambda^2$ and $\log\log Q^2/\log Q^2/\Lambda^2)$ correction terms in the moments (4), which arise from higher order terms in the evolution (1) of $\alpha_s(Q^2)$, in the anomalous dimensions, and in the matrix elements of the quark-antiquark operators. You see that the qualitative features of the fit are not greatly altered, so that one may conclude that QCD perturbation theory is reasonably convergent even at presently accessible values of Q^2. This reflects the fact $Q^2=O(5$ to $10)$ GeV^2 is already quite large on a hadronic scale, as expressed by the value $\Lambda \approx O(500)$ MeV found in typical analyses for the scale parameter Λ appearing in the logarithms (1) and (3) of asymptotic freedom. This sort of scale for the strong interactions means that the typical perturbation parameter $\frac{\alpha_s(Q^2)}{\pi}$ is comfortably small, being O (0.1 to 0.2)

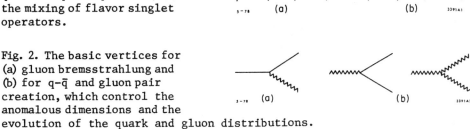

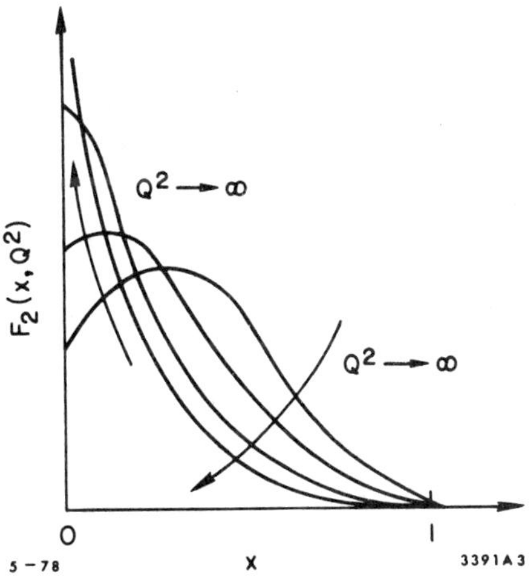

Fig. 3. Qualitative pattern of scaling violation in deep inelastic structure functions anticipated in QCD and other field theories.

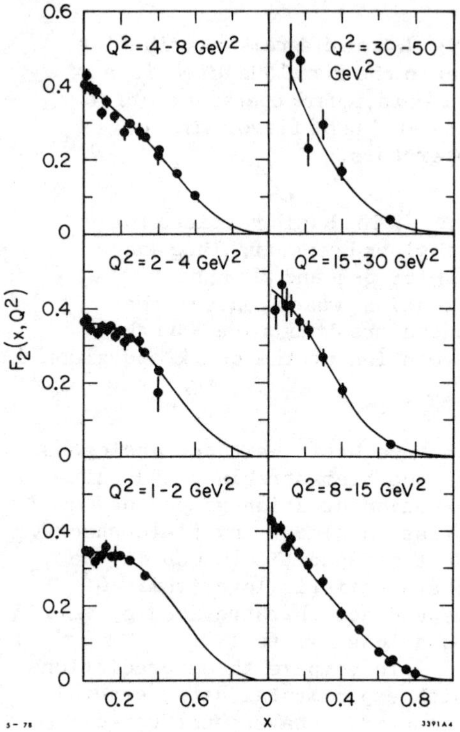

Fig. 4. Data from a FNAL[24] deep inelastic μ scattering experiment.

when $Q^2=0$ (5 to 10) GeV^2.

Although we see from Figs. 4 and 5 that QCD and asymptotic freedom fit the deep inelastic data qualitatively and semi-qualitatively, we would like some more precise numerical vindications of the predictions of the theory. In particular, we would like to have experimental confirmation of the gold-plated predictions (4) for the anomalous dimensions of QCD.[6] A preliminary analysis of just these parameters is now forthcoming from the BEBC group,[7] analyzing νN and $\bar{\nu}N$ charged current data both from their experiment at the CERN SPS and previous Gargamelle data from the CERN PS. They extract from their measured structure functions the moments of the structure functions F_2 and xF_3. One point of sophistication is that they use a modification[29] of the moments (2) which projects on to definite spin even at subasymptotic values of Q^2:

$$M_3(N,Q^2) \equiv \int_0^1 dx\ x^{N-2}\left[\frac{2}{1+\sqrt{1+\frac{4M^2x^2}{Q^2}}}\right]^{N+1} xF_3(x,Q^2)\left[\frac{1+(N+1)\left(1+\frac{4M^2x^2}{Q^2}\right)}{N+2}\right] \tag{7}$$

In formula (7), M is the target nucleon mass, and the modifications to formula (4) just have the effect of removing trivial kinematic dependencies on the mass of the target. The first point to check is that the scaling violations seen in the moments of xF_3 are indeed consistent with the logarithmic Q^2 dependences expected from QCD. This they do by computing the quantities (Fig. 6)

$$[M_3(N,Q)]^{-1/d^N} \tag{8}$$

which should be α log Q^2- log Λ^2. Graphs of the quantities (8) indeed indicate that they are approximately linear in log Q^2,[30] with the same intercept log Λ^2: $\Lambda \sim 700$ MeV, at least for N=2,3,4 and 5. Power dependence in Q^2 with the ratios of anomalous dimensions found in lowest order for a vector gluon theory seems to give a significantly worse fit to the data. The linearity in log Q^2 and the common intercepts of the quantities (8) are non-trivial checks

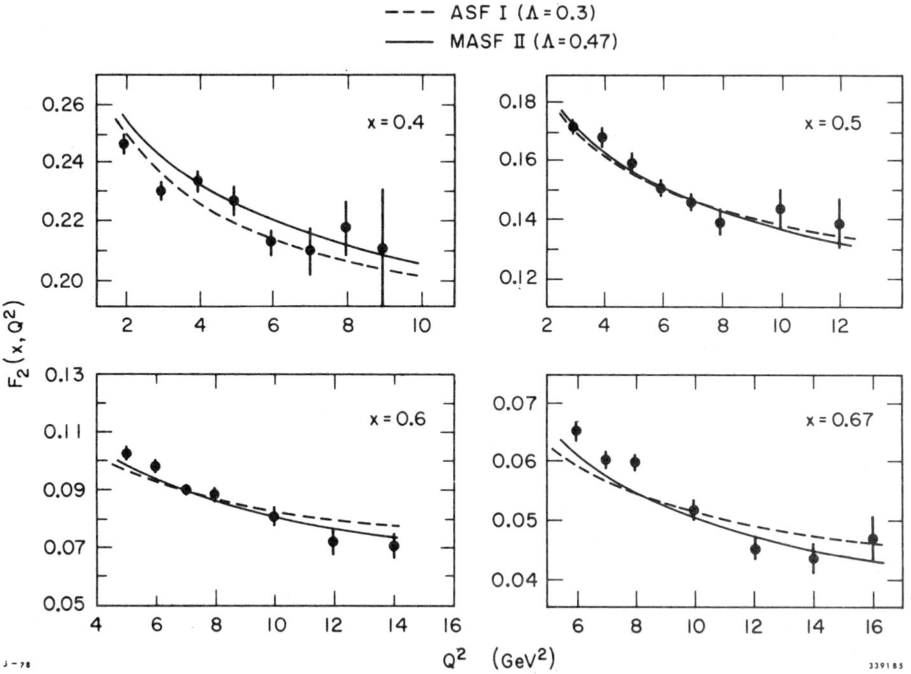

Fig. 5. Typical QCD fit to electroproduction[23] scaling violations, taken from Ref. 27.

of the QCD predictions (4). It is perhaps surprising that these properties hold all the way from Q^2=0(100) GeV^2 all the way down to Q^2=0(1) GeV^2, with no trace of subasymptotic effects due to higher twist operators[31] or quark mass effects.[32] You might also wonder what the effects should be of the higher order QCD perturbation theory effects[28] indicated schematically in equation (3). These have been evaluated by the BEBC group[7] using the results of Ref. 28, and turn out to cause modifications of the linear behaviors of the quantities (8) which are smaller than the present experimental errors for Q^2>0(1) GeV^2. Thus not only do the scaling violations (3) and (4) appear in the BEBC data, but also they are not clearly self-inconsistent.

The most convincing indication of the QCD predictions comes perhaps from comparing the Q^2 dependences of different moments $M_3(N,Q^2)$. Because of the expected asymptotic behaviors (3), the ratios

$$M_3(N,Q^2) / \left[M_3(N',Q^2)\right]^{d_N/d_{N'}} \tag{9}$$

should become constant as $Q^2 \to \infty$. Equivalently, if log $M_3(N,Q^2)$ is plotted against log $M_3(N',Q^2)$ one should see a straight line with slope $d_N/d_{N'}$. Figure 7 shows plots of different combinations of the log $M_3(N,Q^2)$.[7] The solid lines are not fits to the data, but are lines with the slopes predicted by QCD - the agreement is rather good, particularly when you recall that some of the moments vary with increasing Q^2 by about an order of magnitude. The following table[7] is a partial compilation of the best fit slopes on such straight line fits, compared with the theoretical predictions $d^N/d^{N'}$ of QCD. For comparison, the predictions of a model with scalar gluons[33] are also shown.

This remarkable agreement amounts to the first experimental check of unambiguous numbers predicted by QCD. It seems that the characteristic logarithms of QCD are indeed the dominant scale-breaking effects in the xF_3 structure function for Q^2 between 1 and 100 GeV^2. We might term this situation "precocious scaling violation".

In view of the successful comparison of theory and experiment in the xF_3 structure function, it is natural to ask about the F_2 structure function. Here the theoretical situation is more complex because of the new contributions[6] to scaling violations that were mentioned earlier, arising from gluons as well as quarks. It is possible to isolate[7] the gluon piece in QCD taking moments of F_2 and multiplying them by appropriately chosen (precisely specified)[7] functions of Q^2:

$$M_2(N,Q^2)\ Y(Q^2) = M_2(N,Q_o^2) + G(N,Q_o^2)\ X(Q^2) \qquad (10)$$

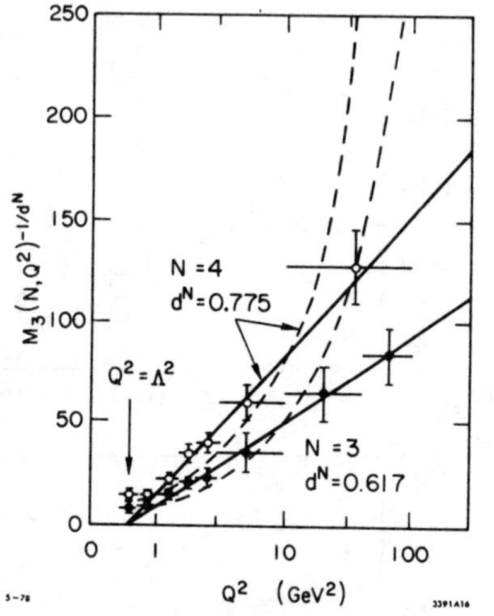

Fig. 6. Nachtmann[29] moments of xF_3, raised to the powers $(-1/d_N)$.[7] QCD predicts an asymptotically linear dependence on log Q^2.

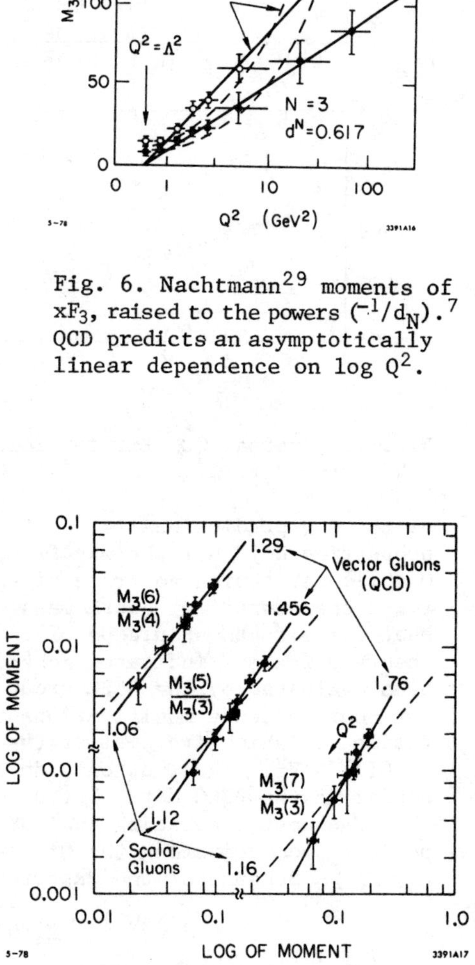

Fig. 7. Plots[7] of the logarithms of moments of xF_3. The solid lines have the slopes $d_N/d_{N'}$ predicted by QCD and other theories with vector gluons. The dashed lines have the slopes predicted by a scalar gluon theory[33] with a small fixed point coupling.

TABLE I

	d_5/d_3	d_7/d_3	d_6/d_2	d_6/d_4
QCD	1.46	1.76	2.53	1.29
Experiment	1.50±0.08	1.84±0.20	3.00±0.51	1.29±0.06
Scalar gluon model	1.12	1.16	1.43	1.06

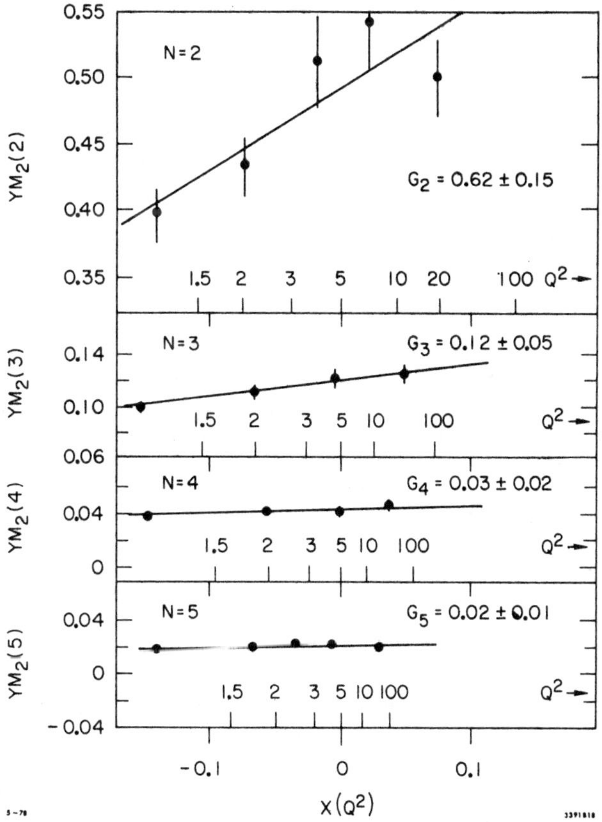

Fig. 8. Scaling violations[7] in the F_2 structure functions. The data would be independent of $X(Q^2)$ if there were no $q\bar{q}$ pair creation from gluons. The solid lines are QCD fits with the gluon moments indicated.

Fig. 9. Radiative corrections to $e^+e^-\to\mu^+\mu^-$ $(q\bar{q})$, (a) with a photon (gluon) in the final state, and (b) with virtual correction to the vertex.

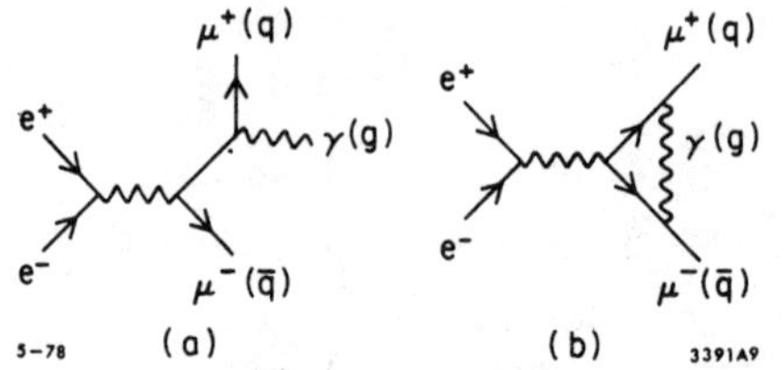

where $G(N,Q_0^2)$ is the N'th moment of the effective gluon distribution $g(x,Q^2)$

$$\begin{matrix} Y(Q^2)\equiv \\ X(Q^2)\equiv \end{matrix} G(N,Q_0^2) = \int_0^1 dx\, x^{N-1} g(x,Q^2) \tag{11}$$

Figure 8 plots a few of the quantities $M_2(N,Q^2)$ $Y(Q^2)$ and indicates the clear need for the second, gluonic term on the right-hand side of equation (10). This term arises from quark-antiquark pair creation in the gluon field of the nucleon. The solid lines correspond to the gluons carrying $^{0}(\frac{1}{2})$ of the nucleon momentum at $Q_0^2=4$ GeV^2,[34] and having an x distribution similar to that of valence quarks. Clearly the data are consistent with these lines, but this analysis is not yet sufficiently advanced to constitute a conclusive test of QCD. A similar analysis of F_2 in deep inelastic eN and μN scattering, retaining all quark and gluon terms, has also been made,[35] with results for $G(2,Q^2)$ and $G(4,Q^2)$ which are compatible with the values used in Fig. 8.

The emerging picture is that not only are the qualitative trends of deep inelastic leptoproduction data[23-25] (see Fig. 4) compatible with QCD, but also semiquantitative QCD fits work very well[26,27] (see Figs. 5 and 8), and furthermore direct experimental quantitative confirmation of the anomalous dimensions predicted by QCD is now becoming available[7] (see Fig. 7 and the Table).[36] We await with interest the forthcoming results of analyses of large statistics CERN-SPS and FNAL νN and μN counter experiments.

3. OTHER APPLICATIONS OF QCD PERTURBATION THEORY

The successes of (renormalization group improved) QCD perturbation theory in the traditional applications to deep inelastic leptoproduction whet our appetite for more areas to apply the theory. There are other applications where the use of asymptotic freedom is underwritten by the renormalization group. But the most dramatic predictions, and the most rapid recent advances, may well lie in areas where the usual machinery of operator product expansions and the renormalization group is not directly relevant. The guiding principle has been that we cannot (yet) disentangle the infrared behavior of QCD, but we are able to study infrared behavior in perturbation theory.[37] The strategy is then to construct experimentally accessible quantities which avoid infrared singularities in perturbation theory, and not obviously vulnerable to incalculable non-perturbative effects. One tactic for doing this is to construct observables totally free of perturbative infrared singularities, such as jet cross-sections in e^+e^- annihilation.[9,38,39,40] Another tactic is to identify a number of processes where the infrared singularities are universal and can be factored out,[41] leaving computable ultraviolet behavior to be compared with experiment. Examples are the Drell-Yan process hadron+hadron→$\ell^+\ell^-$+x,[42] large p_T hadron cross-sections in deep inelastic leptoproduction[43] and hadron-hadron collisions, and final state hadron distributions in leptoproduction and e^+e^- annihilation.[40,44] There follows a brief review of recently developed applications of QCD perturbation theory which adopt one or the other of these two tactics.

Jets in e^+e^- Annihilation

It has been known for a while that the infrared behavior of QCD perturbation theory is generally similar to that of QED.[37] For some time this was a cause for dejection, since it had been hoped that QCD perturbation theory would reveal significant clues to the quark confinement mechanism. But nowadays confinement is expected to be an essentially non-perturbative phenomenon.[12,13] Indeed, we are happy about the infrared similarities between QED and QCD perturbation theory, because they enable us to take over from QED many of the well-understood techniques for constructing quantities which are infrared finite in perturbation theory. In QED it is known[45] that if one introduces any of a range of dimensionless cut-offs - for example demanding that all except a fraction ε of the total center-of-mass energy in some process be emitted in two oppositely directed cones of center-of-mass opening angle δ - then there are no singularities when infrared regulators such as a photon (→gluon) or a lepton (→quark) mass are taken to zero.[46] As an example, consider the lowest order radiative corrections to $e^+e^-\to\mu^+\mu^-$ illustrated in Fig. 9. There are diagrams with real photons in the final state (Fig. 9a), as well as interferences between $e^+e^-\to\mu^+\mu^-$ diagrams with (Fig. 9b) and without virtual radiative corrections. The two classes of diagrams in Figs. 9a and 9b both have infrared singularities, but these cancel if one combines with the pure $\mu^+\mu^-$ final state "degenerate" final states[45] with either a "soft" photon with center-of-mass energy fraction $<\varepsilon$, or a "hard" photon emitted within a cone of angle δ of either the μ^+ or the μ^-. This QED procedure for defining infrared-finite cross-sections has a well-understood extension to all orders of perturbation theory.[45]

Now consider QCD perturbation theory for $e^+e^-\to$ hadrons. The lowest two orders are identical with QED, apart from trivial group-theoretical factors, so that the cancellation of infrared singularities will occur in the same way. Now one interprets[9] the cross-section for all except a fraction ε of the energy to be emitted within two cones of opening angle δ to be the cross-section for $e^+e^-\to$2 jets. The infrared finiteness of the 2-jet cross-section means that

$$\frac{\sigma_{2\ \text{jet}}}{\sigma_{\text{total}}} = 1 - O\left(\frac{\alpha_s}{\pi}\right)$$

$$\frac{1}{\sigma_{\text{total}}}\ \frac{d\sigma(2\ \text{jet})}{d(\cos\theta)} = \frac{3}{4}\left(1+\cos^2\theta\right) + O\left(\frac{\alpha_s}{\pi}\right) \tag{12}$$

where the coefficients of α_s/π in equations (12) depend logarithmically on the dimensionless cut-offs ε and δ, and have recently been computed explicitly.[47] The results (12) hold in QCD perturbation theory, but an act of faith is still necessary to believe that these predictions are not invalidated by non-perturbative effects: perhaps those are connected with the finite p_T width of the jets actually observed[8] in e^+e^- annihilation and elsewhere.

If the infrared similarities between perturbation theory QCD and QED persist in higher orders, then not only two-jet but also multi-jet cross-sections should be predictable in QCD. Indeed, it has recently been shown that infrared singularities vanish for suitably defined multi-jet cross-sections.[48] The diagram of Fig. 9a for gluon bremsstrahlung at wide angles is an embryonic three-jet cross-section with a cross-section[9]

$$\frac{1}{\sigma_{total}} \frac{d^2\sigma(3jets)}{dx_q\, d_{\bar{q}}} = \frac{2\alpha_s}{3\pi} \left[\frac{x_q^2+x_{\bar{q}}^2}{(1-x_q)(1-x_{\bar{q}})}\right] + 0\left(\frac{\alpha_s}{\pi}\right)^2 \qquad (13)$$

where x_q and $x_{\bar{q}}$ are the fractions of the center-of-mass energy Q carried by the quark and antiquark jets respectively: $x_{q(\bar{q})} \equiv 2E_{q(\bar{q})jet}/Q$. We see from equation (13) that the gluon bremsstrahlung three-jet cross-section in the $e^+e^- \to$ hadrons continuum is expected to be $0(\alpha_s/\pi) \overset{?}{\simeq} 0(10)\%$. On the other hand, the dominant decay mode for a heavy 1^{--} quark-antiquark bound state such as the Υ is expected to be to three gluons.[49] Therefore one might expect that for sufficiently massive mesons three-jet final states should predominate,[38] with a cross section

$$\frac{1}{\Gamma_{no\ \gamma^*}} \frac{d^2\Gamma(3jets)}{dx_1\, dx_2} = \frac{1}{\pi^2-9} \left[\frac{(1-x_1)^2}{x_2^2x_3^2} + \frac{(1-x_2)^2}{x_3^2x_1^2} + \frac{(1-x_3)^2}{x_1^2x_2^2}\right] + 0\, \frac{\alpha_s}{\pi} \qquad (14)$$

It would certainly be very nice to verify experimentally the predictions (13) and (14). But how does one look for three-jet final states, particularly if they only constitute a small fraction of the total cross-section?

It has been pointed out[50] that there are directly computable experimental variables which should have distributions free of infrared singularities in QCD perturbation theory and hence be reliably (?) predictable. Final states with exceptional values of these variables should be fertile ground to search for multi-jet final states.[51] The trick is to find variables whose values are identical for the three configurations whose infrared singularities must cancel: a lone quark (Fig. 9b) or a quark and either a hard parallel or a soft gluon (Fig. 9a). Such variables will generally be linear in the momenta. One example is the "thrust" variable[50]

$$T \equiv \max\left[\frac{\sum_{hadrons\ h} |p_{\parallel}^h|}{Q}\right] \qquad (15)$$

where the maximization is performed with respect to the choice of a thrust axis for measuring $p_{\parallel}$. A final state with $T \simeq 1$ will have

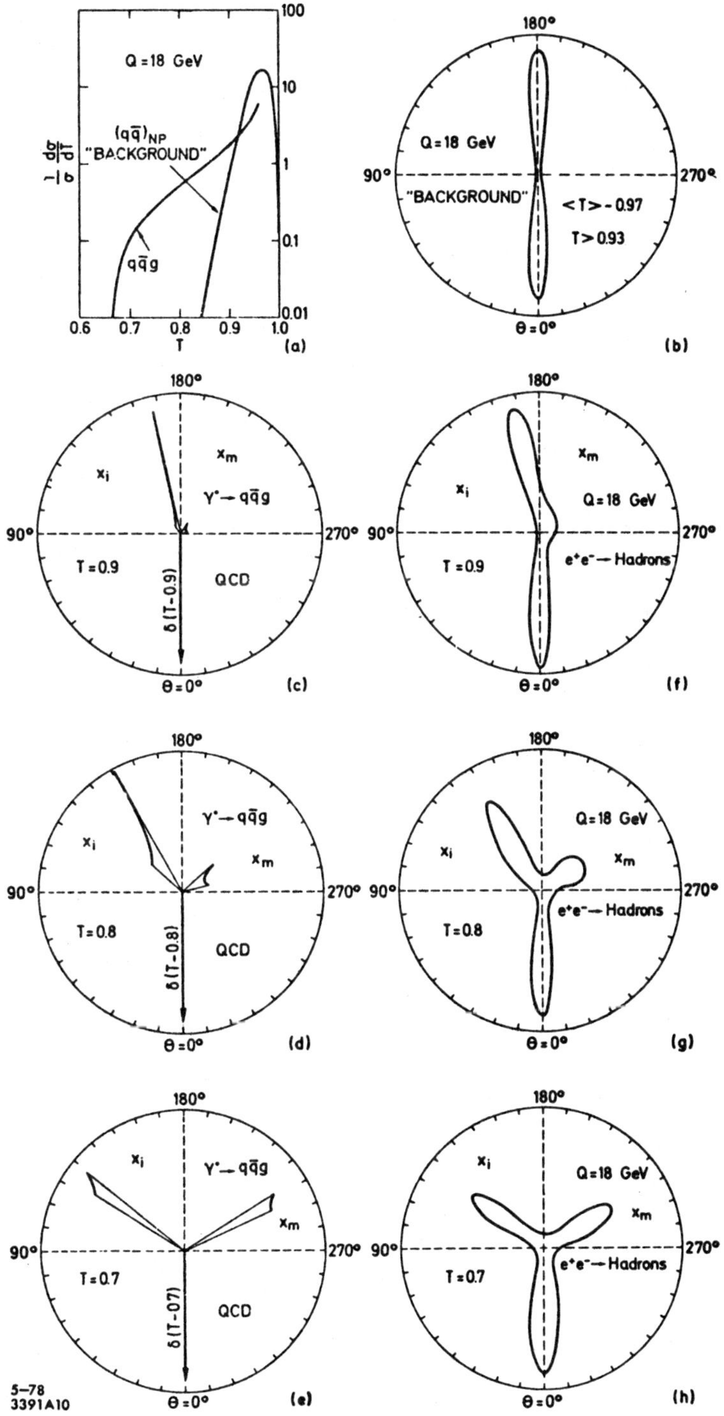

Fig. 10. (a) Comparison[40] of the thrust distribution from QCD perturbation theory with an estimate of the non-perturbative smearing of two-jet configurations (b) at a center-of-mass energy Q=18 GeV, and (c), (d), (e) distributions of the hadron energy (Pointing vector) in the event plane computed in QCD, smeared (f), (g) and (h) by non-perturbative effects.

hadrons with highly collimated momenta, and the final state will have two jets. Events with low thrust T<1 should mostly have a three-jet structure. The putative multi-jet cross-sections (13) and (14) correspond to thrust distributions

$$\frac{1}{\sigma_{total}} \frac{d\sigma}{dT} \approx \frac{2\alpha_s}{3\pi} \left[\frac{2(3T^2-3T+2)}{T(1-T)} \ln \frac{2T-1}{1-T} - \frac{3(3T-2)(2-T)}{(1-T)} \right] \quad (16)$$

and

$$\frac{1}{\Gamma_{no\ \gamma^*}} \frac{d\Gamma}{dT} \approx \frac{3}{\pi^2-9} \left[\frac{4(1-T)}{T^2(2-T)^3} (5T^2-12T+8) \ln \frac{2-2T}{T} + \frac{2(3T-2)(2-T^2)}{T^3(2-T)^2} \right] \quad (17)$$

which are plotted in Figures 10a and 11a respectively. If we select an event with low T, perturbation theory would predict that it be approximately planar. If we plot the angular distribution of the radiated hadronic energy projected on to the event plane (the "Pointing Vector"[40]), then for low T events it should have an angular distribution characteristic of three-jet structures. The results corresponding to (13) and (14), computed with a model for the finite non-perturbative p_T spread of each jet, are shown in Figures 10b to h and 11b to h for the e^+e^- continuum and for Υ decay respectively. If QCD perturbation theory predictions for multi-jet cross-sections are indeed reliable as we believe,[48] then there should be many interesting hadronic final states in e^+e^- annihilation. Similar infrared finite jet predictions can be made for eN, μN, νN and hadron-hadron collisions, but calculating them requires more understanding of infrared singularities associated with individual hadrons in the initial or final state, which we now discuss.

Factorizing Infrared Singularities

Up to now we have permitted our ignorance of the infrared behavior of QCD to restrict ourselves to calculating quantities which have no infrared singularities. But we can relax this criterion by identifying classes of "hard" processes involving large momentum transfers where universal infrared singularities factorize out.[11,41] We can then study ratios of these cross-sections which are only sensitive to the calculable ultraviolet properties of the theory. To see how this program should work, let us contemplate the generic "hard scattering" process illustrated in Fig. 12: a certain number of large Q^2 currents (virtual photons or W bosons?) interact with constituents from a collection of hadrons, of which some are in the initial and some in the final state. We presume that the constituents a, b,... of the hadrons A, B,... have finite (momentum)^2 p_a^2, p_b^2, ..., while the (momentum)^2 Q_i^2 of the currents, and all the momentum transfers between different active participants in the "hard" process all $\to\infty$ in constant ratios. Suppose we calculate in QCD perturbation theory using a coupling constant renormalized at some large momentum Q_0^2: $\alpha_s(Q_0^2)$. Then we will encounter logarithms in the relevant Feynman diagrams which are of two types: $\ln(Q^2/Q_0^2)$ (to be called ultraviolet), and $\ln(Q^2/p^2_{a,b,\ldots})$ (to be called infrared).

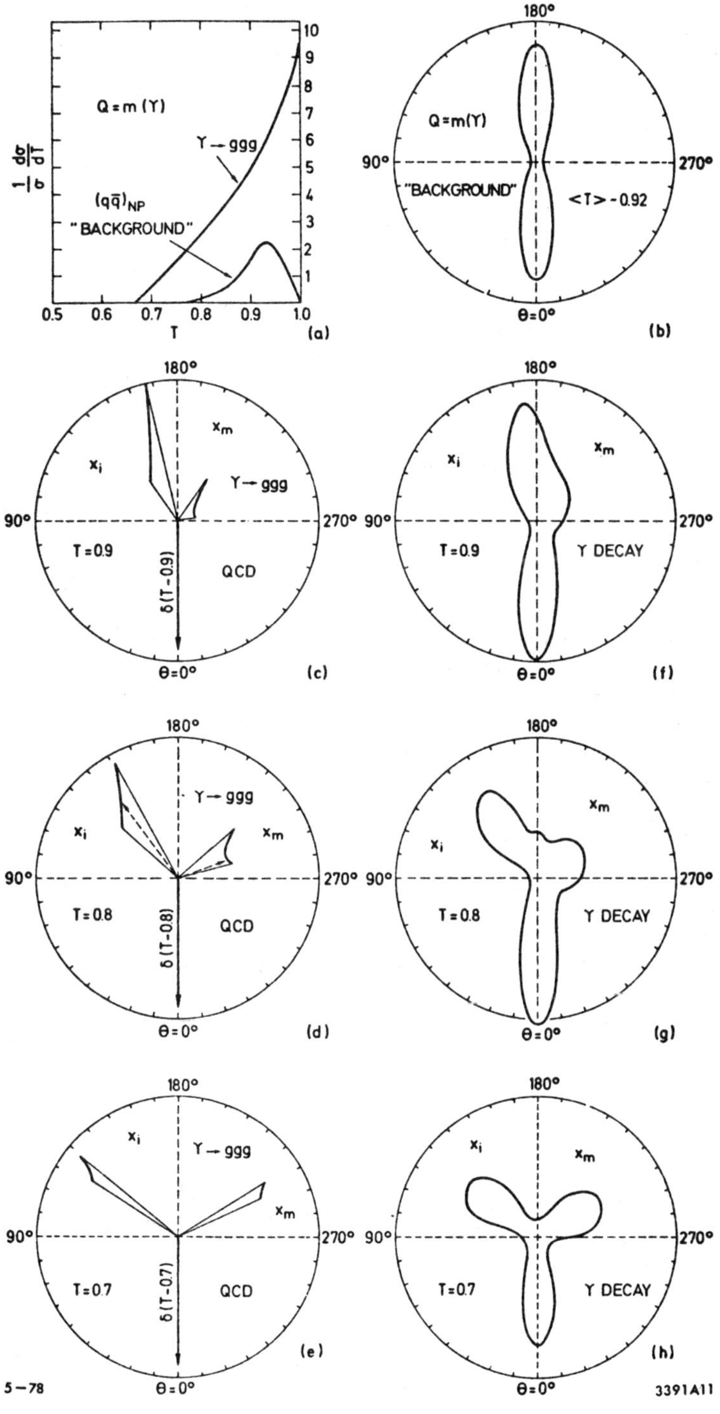

Fig. 11. Similar to Fig. 10, computed[40] for Υ → 3 gluons.

The only dependence on the bound state properties resides in the infrared logarithms, and the hope is that this dependence factorizes[11,41] into separate universal terms for each external hadronic leg[52]:

$$\sigma \approx \sigma_{Hard}(\alpha_s(Q_o^2), \ln Q^2/Q_o^2) \times \prod_{A,B,\dots} F_{Aa}(\alpha_s(Q_o^2), \ln Q^2/Q_o^2, \ln Q^2/p_a^2) \quad (18)$$

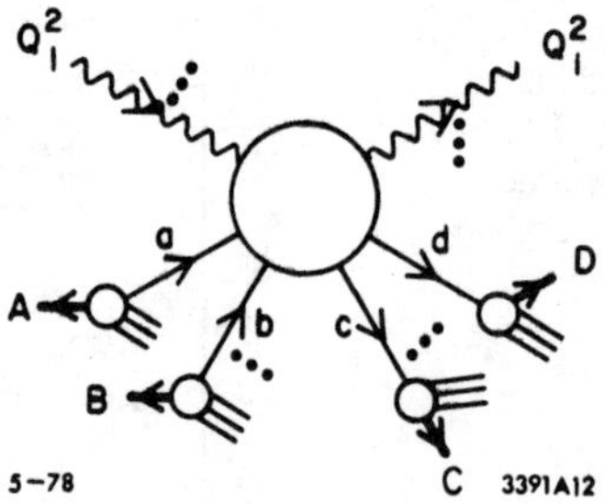

Fig. 12. A generic hard scattering diagram with deep inelastic currents Q_1^2 to Q_i^2, and constituents a, b, c and d of hadrons A, B, C and D. All momentum transfers and Q^2 are supposed large.

The ultraviolet logarithms will then arrange themselves so that the cross-section can be re-expressed in terms of the coupling constant at Q^2: $\alpha_s(Q^2)$, and one can then rewrite

$$\sigma \approx \sigma_{Hard}(\alpha_s(Q^2)) \times \prod_{A,B,\dots} F_{Aa}(Q^2)$$

where σ_{Hard} will be given in leading order by the Born terms. The universal distribution functions $F_{Aa}(Q^2)$ will be the same in different scattering processes. In particular, those connected with initial state hadrons will be the same as in deep inelastic leptoproduction,[41] and so should be identified as the effective quark and gluon distributions introduced earlier and obeying[22] equations like (5). If there is also factorization of the infrared logarithms relating to final state hadrons,[11] then one would also have universal quark (or gluon)→hadron fragmentation functions, which would play roles analogous to those in the naive parton model,[4] albeit with scaling violations analogous to those for the initial distribution functions (cf equations (2) to (5)). If we want, we may sum over all hadrons emanating from one of the final state constituents, in which case we expect to cancel the associated infrared singularities and arrive at an infrared finite and hence calculable jet cross-section.

To what "hard" scattering reactions can this approach be applied, and what calculations have been done to support the factorization picture outlined above? If we first look at the total electroproduction cross-section <u>ep→e+x</u> illustrated in Fig. 13 then the derived

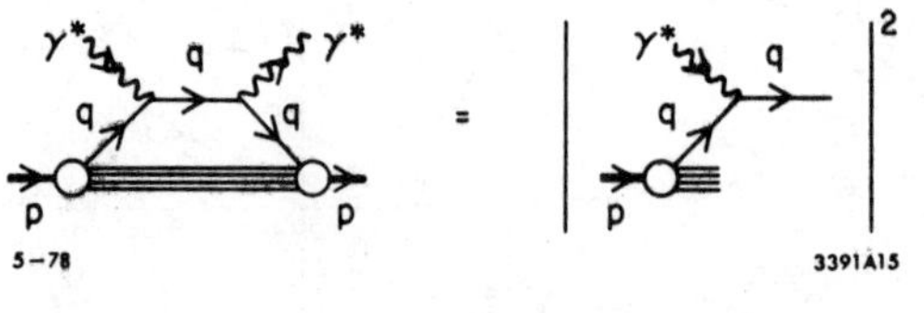

Fig. 13. A sample "hard" scattering cross-section: the total electroproduction cross-section is the modulus squared of the hard subprocess $\gamma^*+q\to q$, folded with the distribution $q(x,Q^2)$ of quarks in the proton.

factorization is guaranteed by the renormalization group.[1,41] It tells us that the development of the structure function at large Q^2

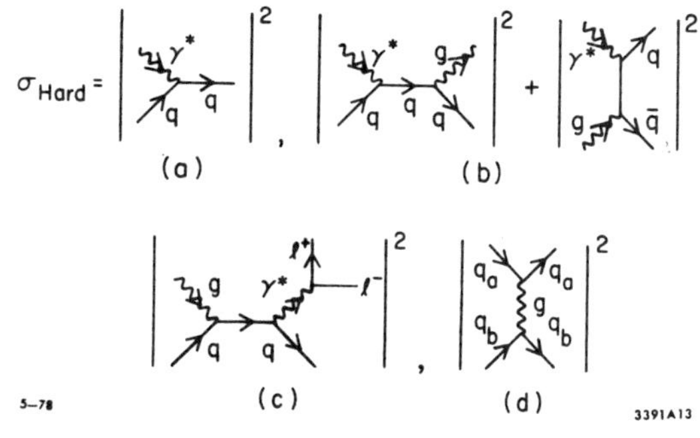

Fig. 14. Hard subprocesses relevant to (a) the electroproduction cross-section, (b) large p_T production in electroproduction; (c) Drell-Yan pair production at large p_T and (d) large p_T production in hadron-hadron collisions.

is controlled by the anomalous dimensions (4), or equivalently the evolution equation (5). The Born term for σ_{Hard} in equation (19) is just the point-like charge coupling of Fig. 14a. The infrared sensitivity resides in the boundary conditions to be fed into the evolution equation (5). When the leading logarithms in $e^+e^- \to$ hadron C+x are studied,[11,43,53] they are also found to exhibit the desired factorization, with the point-like Born term of Fig. 14a and an infrared sensitive quark→hadron fragmentation function $F_{c\to C}(Z,Q^2)$, where $Z \equiv E_C/E_c$. The high Q^2 development of $F_{c\to C}$ is controlled by anomalous dimensions and an evolution equation which is just the analytic continuation of equation (5), and reflects the same physical processes of bremsstrahlung and pair creation. (Notice though that the initial conditions for these evolution equations are infrared sensitive and not related by analytic continuation.) The factorization property becomes very important when it is demonstrated and used for e+p→e+C+x, because it means that a simple partonesque[4] "building block" formula applies to the cross-section for producing final state particles with longitudinal momentum fraction Z at low p_T:

$$\sigma(e+p\to e+C+x = \sum_{q,\bar{q}} q(x,Q^2)e_q^2\, F_{q\to C}(Z,Q^2) \qquad (20)$$

A non-partonesque[4] process is the production of large p_T jets in electroproduction.[43] Here the lowest order hard process is wide angle "Compton scattering" $\gamma^*+q\to g+q$ (or pair production $\gamma^*+g\to q+\bar{q}$) as in Fig. 14b, and if the factorization property holds the only infrared singularities are those connected with the target proton, which are the same as in the total electroproduction cross-section, so that one may write

$$\sigma_{\text{large } p_T \text{ jets}} = \sum_{q,\bar{q}} q(x,Q^2)\,\sigma(\gamma^*+q\to g+Q))\Big|_{\alpha_s(Q^2)} + \begin{matrix}\text{gluon}\\ \text{terms}\end{matrix} \qquad (21)$$

Of course one may always study the production of individual final state hadrons at large p_T, in which case formula (21) is just convoluted with the same fragmentation function $F_{q\to C}$ (Z,Q^2) as appeared in equation (20). And so it goes.

So far we have not considered processes with two initial state hadrons: the simplest such reaction is the Drell-Yan process $pp\to\ell^+\ell^-+x$. Here again, infrared factorization[42] has been demonstrated[53] to all orders at the leading logarithm level, meaning that the naive

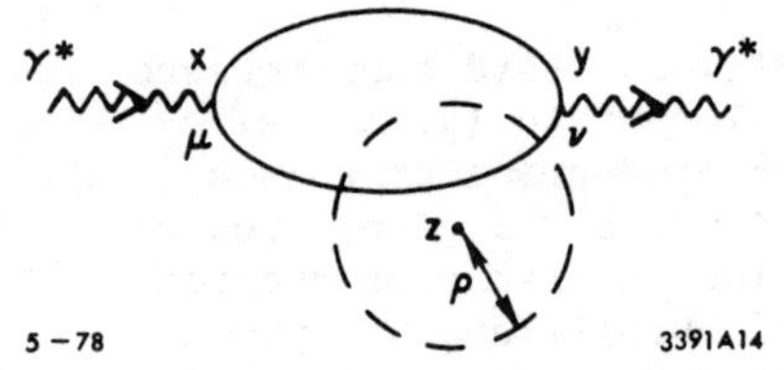

Fig. 15. The single fermion loop calculated[14] in the presence of a background instanton field of size ρ located at the point z.

QCD modification of the usual parton cross-section formula is indicated:

$$M^4\frac{d\sigma}{dM^2} = \frac{4\pi\alpha^2}{9}\sum_{q\bar{q}}\int_0^1 dx_q \int_0^1 dx_{\bar{q}}\; e_q^2\, q\,(x_q, M^2)\,\bar{q}\,(x_{\bar{q}}, M^2)\,\delta(x_q x_{\bar{q}} - M^2/s) + 0(\alpha_s/\pi) \qquad (22)$$

There are however some important QCD corrections to the usual parton physics of this reaction. First, the essential scale invariance of QCD at short distances means that there is no p_T cut-off, and therefore the average p_T^2 of the lepton pair should grow $\alpha\, M^2$ (to within logarithms). Care should be taken in comparing this prediction with existing data, since $\langle p_T^2\rangle$ may also depend on the value of $\tau \equiv M^2/s$ being studied.[54] The dependence on M^2 and τ should be carefully disentangled when comparing theory and experiment, but the growth with M^2 at fixed τ does seem to be present in the data. Another important remark concerns the $0(\alpha_s/\pi)$ pieces in equation (22): the basic $0(\alpha_s/\pi)^0$ $q\bar{q}$ annihilation piece may be rather small in pp collisions, just because the proton contains relatively few antiquarks. Rival hard subprocesses which are naively suppressed by $0(\alpha_s/\pi)$ such as $g+q\to(\ell^+\ell^-)+q$ (see Fig. 14c), or even $0(\alpha_s/\pi)^2$ like $q+q\to(\ell^+\ell^-)+q+q$, may also be just as important phenomenologically.[55] The Drell-Yan process is probably rather free of these problems in πp or $\bar{p}$p collisions: their significance for the total $pp\to(\ell^+\ell^-)+x$ cross-section has not yet been fully calculated.

The final "hard" process we should mention is pp→large p_T hadrons + x. This process is interesting because it may constitute the first theoretically "clean" testing-ground for QCD in pure hadronic collisions. The required factorization of all initial- and final-state infrared logarithms has been verified for the leading logarithms in the lowest non-trivial order of perturbation theory,[56] and arguments exist that this factorization persists in higher orders.[57] If so, this means that at very large transverse momenta the correct QCD prescription for the quark-quark cross-section of Fig. 13d is

$$E_C\frac{d\sigma}{d^3p_C}(A+B\to C+x) = \sum_{a,b,c,d}\int_0^1 dx_a \int_0^1 dx_b \int_0^1 \frac{dx_c}{x_c^2} \times \left[q_a(x_a,s')\,q_b(x_b,s')\,F_{c\to C}(x_c,s')\right] \times \delta(s'+t'+u')\,\frac{s'}{\pi}\,\frac{d\sigma}{dt'}(q_a+q_b\to q_c+q_d)\Big|_{\alpha_s(s')} + \ldots \qquad (23)$$

where the dots comprise the "hard" scattering of other types of constituent, terms of higher order in α_s/π, and so on. The QCD large p_T formula (23) has many sources of scaling violation, and its application is complicated by the non-zero p_T of the initial-state constituents. The writer is no expert on large p_T phenomenology, but it does seem possible that some fraction of present large p_T data may indeed result from the fundamental "hard" scattering diagrams of QCD.

Recent theoretical progress has made available many predictions of QCD perturbation theory for "hard" processes beyond the traditional applications of Section 2 which were guaranteed by operator product expansions and the renormalization group. It would be important to find a comparably reliable and general theorem to underpin these new applications of QCD perturbation theory, but already they look to be reliable predictions eager to be confronted with experiment.

4. NON-PERTURBATIVE EFFECTS AT LARGE MOMENTA

So far we have concentrated on results which hold in perturbation theory in QCD. But we believe that non-perturbative effects[12] have some very important phenomenological consequences, such as quark confinement.[13] We should ask whether non-perturbative effects are also important at large momenta, where we have made our applications of QCD perturbation theory so far. The normal procedure for studying non-perturbative phenomena is to use the functional integral and construct semi-classical approximations to it. The WKB approach first looks for solutions of the classical Euclidean field equations,[12] which correspond to minima of the action and hence stationary points of the functional integrand. One then does perturbation theory around each classical solution, adds them together weighted by the classical Euclidean actions e^{-A}, and finally continues amplitudes back to Minkowski space.[58] This procedure is incomplete, in that there may be other non-perturbative configurations with larger actions, which nevertheless occur in large enough numbers (entropy) to overcome the exponential suppressions of their e^{-A} factors.[13] As a starting point we ignore this complication and concentrate on the classical extrema, which are known in the case of QCD to be multi-instanton (-anti-instanton) configurations.[12] The action of a single instanton is known to be

$$A = \frac{8\pi^2}{g^2} \tag{24}$$

where one-loop perturbative calculations[59] in the presence of an instanton indicate that g^2 in equation (24) is to be interpreted as $g^2(\rho)$ (of equation (1) with $Q^2 \approx \frac{1}{\rho^2}$), where ρ is the usual size parameter of the instanton. Asymptotic freedom (1) means that $A\to\infty$ when the size $\rho\to 0$, so that small instantons make relatively small contributions to the functional integral. We might hope that processes at large Q^2 would somehow only "feel" small instantons of size $\rho\sim 1/Q$, in which case the weighing of e^{-A} and equations (1) and (24) would suggest that non-perturbative effects would fall like a power of Q^2 as $Q^2\to\infty$.

Some calculations to explore this possibility have been done recently[14]: what has been studied is the contribution of the simplest non-trivial non-perturbative configuration (one instanton) to the most basic short distance or deep inelastic process ($e^+e^-\to$hadrons at high Q^2), evaluating the lowest order Feynman diagram (the simple fermion loop of Fig. 14). If the contribution to $\sigma(e^+e^-\to\text{hadrons})$ at large Q^2 were dominated by small instantons, then this calculation could be regarded as the leading term of a non-perturbative calculation in the conventional "dilute gas approximation"[13] in which instantons (and anti-instantons) are supposed to be non-interacting and not very dense.

It does indeed turn out that this one-instanton contribution to $R\equiv\sigma(e^+e^-\to\text{hadrons})/\sigma(e^+e^-\to\mu^+\mu^-)$ is controlled by such small instantons of size $\rho=0(^1/Q)$. The fermion loop of Fig. 13 is easily calculated[14] using the known[60] fermion propagator $S^1(x,y;z,\rho;m)$ in the presence of an instanton located at z and of size $\rho>>^1/m$:

$$\delta\pi_{\mu\nu}(x,y)\equiv\langle 0|T(J_\mu(x)J_\nu(y))|0\rangle_{\text{instanton}} = \int d^4z_\mu \int_0^\infty d\rho\, d(\rho) \Big[-\text{Tr}(\gamma_\mu S^1(x,y;z,\rho;m)\gamma_\nu S^1(y,x;z,\rho;m)) + \text{Tr}(\gamma_\mu S^o(x,y;m)\gamma_\nu S^o(y,x;m))\Big] \quad (25)$$

The density function $d(\rho)$ for the instantons is known from the work

$$d(\rho) = \frac{0\cdot 1}{\rho^5}\left(\prod_{i=1}^{N_F} m_i\right)\left(1\cdot 3\ \Lambda\rho\right)^{N_f} \exp\left(\frac{-8\pi^2}{g^2(\rho)}\right)\left(\frac{8\pi^2}{g^2(\rho)}\right)^6 \quad (26)$$

Transforming equation (25) to momentum space one finds

$$\delta\Pi^\mu_\mu(Q) = \int_0^\infty d\rho\, d(\rho)\left[\frac{-12}{Q^2} + 36\rho^2\int_0^1 dx K_2\left(\frac{2\rho Q}{\sqrt{1-u^2}}\right)\right] \quad (27)$$

The first term in equation (27) gives a contribution to the real part of the vacuum polarization which is suppressed by $0(^1/Q^4 \ln Q^2)$ relative to the leading perturbative contribution.[14] However its (singular) contribution to the absorptive part relevant to R is cancelled by the singularity in the second term at $Q^2=0$. The only contribution to the instanton piece (27) explicitly comes[14] from $\rho=0(^1/Q)$, and yields

$$\frac{\Delta R}{R}\underset{\substack{Q\to\infty\\ N_f\neq 3,9,\ldots}}{\approx} \left(\frac{Q}{\Lambda}\right)^{-11-N_f/3} \left(\ln Q^2\right)^{6\left(\frac{33-4N_f}{33-2N_f}\right)} \times 6\pi^{3/2}(0\cdot 1)(1\cdot 3)^{N_f}\left(\frac{33-2N_f}{6}\right)^6 \cos\frac{\Pi N_f}{6} \times \frac{\Gamma\left(\frac{7}{2}+\frac{N_f}{6}\right)\times\left[\Gamma\left(\frac{11}{2}+\frac{N_f}{6}\right)\right]^2}{\Gamma\left(6+\frac{N_f}{6}\right)}\left(\frac{\prod_{i=1}^{N_f}\bar{m}_i}{\Lambda^{N_f}}\right) \quad (28)$$

with a related form[14] for $N_f = 3,9,\ldots$

A numerical evaluation relevant between the strangeness and charm threshold yields

$$\frac{\Delta R}{R} \approx \left(\frac{Q}{1{\cdot}5\ \mathrm{GeV}}\right)^{-12} (\ln Q^2)^{3{\cdot}67} \tag{29}$$

if $\Lambda \sim 500$ MeV. The detailed form and value of ΔR (28) and (29) are probably phenomenologically meaningless and undetectable, but they indicate that in e^+e^- annihilation there is probably some sort of non-perturbative "brick wall" somewhere between 1 and 2 GeV in the center-of-mass. Above this point QCD perturbation theory probably reigns supreme, whereas at lower energies non-perturbative effects are O(1). It is encouraging that this is just the region where vector meson resonances are known to lie.

It is interesting that this simplest possible calculation supports the "non-perturbative perturbation theory" picture we hoped to find. But many more calculations are needed of more complex diagrams for more complicated processes in the presence of more intricate background fields before we can feel sure that non-perturbative phenomena are always small at large Q^2. Nevertheless, we may legitimately hope that non-perturbative effects will not invalidate the elegant pattern of perturbative QCD predictions that we discussed in Sections 2 and 3.

5. CONCLUSIONS AND QUESTIONS

On the basis of the previous discussion, the following seem reasonable conclusions about the status of phenomenological applications of QCD:

The classic asymptotic freedom predictions of renormalization group improved QCD perturbation theory for deep inelastic leptoproduction triumphantly pass present tests. In addition to the familiar qualitative successes of QCD fits to deep inelastic scaling violations, we now find for the first time that the anomalous dimensions predicted by QCD are receiving direct experimental confirmation. We look forward to detailed analyses of the large statistics νN and μN experiments which should be available soon. In particular we may hope for analyses of the flavor singlet structure functions which are as striking as the BEBC analysis[7] of the xF_3 structure function.

Recent theoretical work suggests that there are many other high momentum transfer processes where QCD perturbation theory can legitimately be applied to make reliable predictions. These include jet production, final state hadron distributions in deep inelastic processes at small and large p_T, Drell-Yan lepton pair production and hadron-hadron collisions at large p_T. We look forward to experimental tests of these predictions.

Finally, we see reason to hope that non-perturbative QCD effects may not invalidate the predictions of QCD perturbation theory that we have been discussing. Unfortunately, the other side of this coin is that a convincingly calculable and testable phenomenological manifestation of non-perturbative QCD has yet to be identified.

This is just one of many open questions about applications of

QCD. A partial list might include: How can we be sure the strong coupling constant is really asymptotically free? Many of the tests of Section 2 would be successful if there was a fixed point reasonably close to the origin, so that moments violated scaling by small powers of Q^2 instead of log Q^2. Presumably we should do experiments at much higher Q^2, using for example[30] high energy electron-proton colliding rings? What about σ_L/σ_T, where experimental results are not in dramatic agreement with experiment?

Presently observed jets seem to have a finite p_T cut-off. Can we understand this, or (which may be equivalent) be sure that non-perturbative effects really don't mess up our perturbative jet predictions? Can we say something about the hadronic wavefunctions or fragmentation functions? At the moment we just factor out our ignorance about them. Although we have not discussed them here, heavy quark-onium states seem particularly amenable to perturbative analysis, but we are still short on rigorous theoretical understanding.

Finally, can we find some unambiguous and calculable physical manifestation of the non-perturbative QCD phenomena now being intensively studied? We hope ultimately to be able to calculate the hadron spectrum, but in the meantime is there any way of "seeing" an instanton? Or a meron?

There are still more open questions than closed solutions.

ACKNOWLEDGMENTS

It is a pleasure to thank J.D. Bjorken, S.J. Brodsky and R.C. Brower for useful comments and discussions, and S.D. Drell for the hospitality of the SLAC theory group.

REFERENCES AND FOOTNOTES

1. For general reviews of QCD, see H.D. Politzer, Phys. Reports 14C, 129 (1974); W. Marciano and H. Pagels, Phys. Reports 36C, 137 (1977).
2. D.J. Gross and F.A. Wilczek, Phys. Rev. Lett. 30, 1343 (1973); H.D. Politzer, Phys. Rev. Lett. 30, 1346 (1973).
3. T. Applequist and H.D. Politzer, Phys. Rev. Lett. 34, 43 (1975) and Phys. Rev. D12, 1404 (1975). For a recent phenomenological review and references see K. Gottfried, Proceedings of the 1977 International Symposium on Lepton and Photon Interactions at High Energies, ed. F. Gutbrod, (DESY, 1977) p. 667.
4. J.D. Bjorken and E.A. Paschos, Phys. Rev. 158, 1975 (1969); S.D. Drell, D.J. Levy and T.-M. Yan, Phys. Rev. D1, 1035 (1970); R.P. Feynman, "Photon-Hadron Interactions" (Benjamin, New York, 1972).
5. S. Okubo, Phys. Lett. 5, 165 (1963); G. Zweig, CERN preprints TH 401, 412 (1964); J. Iizuka, Suppl. Progr. Theor. Phys. 37-38, 21 (1966). For other discussions of QCD and "soft" hadronic processes, see G. Veneziano, Nucl. Phys. B117, 519 (1976) and references therein.
6. D.J. Gross and F.A. Wilczek, Phys. Rev. D8, 3633 (1973) and D9, 980 (1974); H. Georgi and H.D. Politzer, Phys. Rev. D9, 416 (1974).

7. P.C. Bosetti et al., CERN preprint "Analysis of Nucleon Structure Functions in CERN Bubble Chamber Neutrino Experiments" (1978). For a preliminary analysis, see K. Schultze, Proceedings of the 1977 International Symposium on Lepton and Photon Interactions at High Energies, Hamburg, ed. F. Gutbrod (DESY, 1977), p. 359. I thank Bill Scott for useful discussions about the analysis.

8. G. Hanson et al., Phys. Rev. Lett. 35, 1609 (1975).

9. For an early theoretical discussion of jets, see A.M. Polyakov, Soviet Phys. JETP 32, 296 (1971); 33, 850 (1971). For the more recent application to QCD, see J. Ellis, M.K. Gaillard and G.G. Ross, Nucl. Phys. B111, 253 (1976); G. Sterman and S. Weinberg, Phys. Rev. Lett. 39, 1436 (1977).

10. S.D. Drell and T.-M. Yan, Phys. Rev. Lett. 25, 316 (1970). For the most complete analysis in QCD perturbation theory see Yu. L. Dokshitzer, D.I. D'yakonov and S.I. Troyan, Materials for the XIIIth Leningrad Winter School (1978), p. 3 and Leningrad Nuclear Physics Institute preprint "Hard Semi-inclusive Processes in QCD" (1978).

11. For early discussions see A. Mueller, Phys. Rev. D9, 963 (1974); C. Callan and M. Goldberger, Phys. Rev. D11, 1553 (1975). There is a heuristic review by A.M. Polyakov, Proceedings of the 1975 International Symposium on Lepton and Photon Interactions at High Energies, Stanford, ed. W.T. Kirk (SLAC, 1975) p. 855. For a discussion in the context of QCD see H. Georgi and H.D. Politzer, Harvard University preprint HUTP-77/A071 (CALT-68-629)(1977).

12. A.A. Belavin, A.M. Polyakov, A.S. Schwartz and Yu.S. Tyupkin, Phys. Lett. 59B, 85 (1975). For a complete review, see S. Coleman, Harvard University preprint HUTP-78/A004 (1978).

13. A.M. Polyakov, Nucl. Phys. B121, 429 (1977); C. Callan, R. Dashen and D.J. Gross, Princeton Institute for Advanced Study preprint COO-2220-115 (1977).

14. N. Andrei and D.J. Gross, Princeton University preprint "The Effect of Instantons on the Short Distance Structure of Hadronic Currents" (1978); R.D. Carlitz and C. Lee, Pittsburgh University preprint PITT-193 (1978); L. Baulieu, J. Ellis, M.K. Gaillard and W.J. Zakrzewski, CERN preprint TH 2482 (1978).

15. Although few theorists would seriously question the necessity of discussing non-perturbative phenomena in QCD, and believe that the functional integral is the correct framework for studying them, I am unaware of any unassailable theoretical or phenomenological proof that they must be incorporated in the fashion of references 12, 13 and 14. This makes non-perturbative effects all the more fascinating.

16. N_f is the number of quark flavors which have masses $m_i << Q$, and so can be presumed to have "switched on" at the momenta of interest. See H. Georgi and H.D. Politzer, Phys. Rev. D14, 1829 (1976). Λ is a mass parameter setting the scale of the strong interactions which is arbitrary a priori, but generally expected to be O(300 MeV to 1 GeV).

17. J.M. Cornwall and R.E. Norton, Phys. Rev. 177, 2584 (1969). The variable x is the conventional Bjorken scaling variable $x=Q^2/2\nu$.

Asymptotically at large Q^2 these moments project on to channels with definite spin N in the crossed channel. There are modifications to the Cornwall-Norton moments which make precise spin projections at finite Q^2: see equation (7) and reference 29.

18. For a negative search for asymptotic freedom in non-gauge theories, see S. Coleman and D.J. Gross, Phys. Rev. Lett. 31, 851 (1973); though it is possible to construct implausible spontaneously broken gauge theories with asymptotic freedom. The renormalization group says that theories with ultraviolet fixed points away from the origin will violating scaling by powers of Q^2: K.G. Wilson, Phys. Rev. 179, 1499 (1969).
19. N. Christ, B. Hasslacher and A. Mueller, Phys. Rev. D6, 3543 (1972).
20. For a very physical discussion, see J. Kogut and L. Susskind, Phys. Rev. D9, 697 and 3391 (1974).
21. G. Parisi, Phys. Lett. 43B, 207 (1973); D.J. Gross, Phys. Rev. Lett. 32, 1071 (1974).
22. G. Altarelli and G. Parisi, Nucl. Phys. B126, 298 (1977).
23. R.E. Taylor, Proceedings of the 1975 International Symposium on Lepton and Photon Interactions at High Energies, Stanford, ed. W.T. Kirk (SLAC, 1975) p. 679; E.M. Riordan et al., SLAC preprint PUB-1634 (1975).
24. L.N. Hand, Proceedings of the 1977 International Symposium on Lepton and Photon Interactions at High Energies, Hamburg, ed. F. Gutbrod (DESY, 1977) p. 417. The data shown here come from the Chicago-Harvard-Illinois-Oxford group, courtesy of T.W. Quirk.
25. For some published data see ref. 7 and T.H. Burnett, Proceedings of the 1977 International Symposium on Lepton and Photon Interactions at High Energies, Hamburg, ed. F. Gutbrod (DESY, 1977) p. 227. Similar trends are apparent in the data of the CDHS counter experiment at CERN.
26. For a review see O. Nachtmann, Proceedings of the 1977 International Symposium on Lepton and Photon Interactions at High Energies, Hamburg, ed. F. Gutbrod (DESY, 1977), p. 811.
27. A.J. Buras, E.G. Floratos, D.A. Ross and C.T. Sachrajda, Nucl. Phys. 131, 308 (1977).
28. E.G. Floratos, D.A. Ross and C.T. Sachrajda, Nucl. Phys. B129, 66 (1977). There are some minor errors in these calculations which do not seem to affect the qualitative phenomenological conclusions. See W.A. Bardeen, A.J. Buras, D. Duke, and T. Muta, FNAL preprint 78/42 THY (1978); also E.G. Floratos, D.A. Ross and C.T. Sachrajda, private communication.
29. O. Nachtmann, Nucl. Phys. B63, 237 (1973) and Nucl. Phys. B78, 455 (1974), S. Wandzura, Nucl. Phys. B122, 412 (1977).
30. To be really sure of this, we need the largest feasible lever arm in Q^2. A good way to realize this is with high energy e-p colliding rings, see for example "CHEEP - an e-p facility in the SPS", CERN Yellow Report 78-02, edited by J. Ellis, B.H. Wiik and K. Hübner (1978).
31. The possibility that high twist operators might give phenomenologically significant scaling violations was discussed in particular by I.A. Schmidt and R. Blankenbecler, Phys. Rev. D16,

1318 (1977).

32. See Georgi and Politzer, ref. 16 and A. De Rújula, H. Georgi and H.D. Politzer, Ann. Phys. 103, 315 (1977); R. Barbieri, J. Ellis, M.K. Gaillard and G.G. Ross, Nucl. Phys. B117, 50 (1977).

33. See D. Bailin and A. Love, Nucl. Phys. B75, 159 (1974); M. Glück and E. Reya, Phys. Rev. D16, 3242 (1977). Abelian vector gluon theories give the same non-singlet anomalous dimensions as QCD in order g^2. Therefore the Table and Figure 7 should best be regarded as evidence for the vector nature of the gluons. The main evidence that the correct theory is QCD rather than an (abelian or non-abelian) vector gluon theory with a non-zero ultra-violet fixed point coupling is the logarithmic Q^2 dependence of the moments discussed earlier.

34. See for example D.H. Perkins, Proceedings of the 16th International Conference on High Energy Physics, Chicago-Batavia 1972 (NAL, Batavia, 1972) p. 189. The fraction is Q^2 dependent: for a more sophisticated evaluation see H.D. Politzer, Nucl. Phys. B122, 237 (1977).

35. H.L. Anderson, H.S. Matis and L.C. Myrianthopoulos, Phys. Rev. Lett. 40, 1061 (1978).

36. The only cloud on the deep inelastic horizon is the apparent discrepancy between QCD calculations (see ref. 32) and measurements of $R=\sigma_L/\sigma_T$ in electroproduction - see L.N. Hand, ref. 24. We hope that this disagreement arises from the systematic problems in making the measurements of R, but we should avoid dogmatism.

37. See for example J. Frenkel, R. Meuldermans, I. Mohammed and J.C. Taylor, Nucl. Phys. B121, 58 (1977); and Section 5 of Marciano and Pagels, ref. 1.

38. T.A. DeGrand, Y.J. Ng and S.-H.H. Tye, Phys. Rev. D16, 3251 (1977).

39. K. Koller and T.F. Walsh, Phys. Letters 72B, 227 (1977), erratum (to be published) and DESY preprint DESY 78/16 (1978); M. Krammer and H. Krasemann, Phys. Lett. 73B, 58 (1978); S.J. Brodsky, D.G. Coyne, T.A. DeGrand and R.R. Horgan, Phys. Lett. 73B, 203 (1978); H. Fritzsch and K.-H. Streng, Phys. Lett. 74B, 90 (1978).

40. A. De Rújula, J. Ellis, E.G. Floratos and M.K. Gaillard, CERN preprint TH 2455 (1978).

41. While foreshadowed in the first two papers of ref. 11, the present surge of interest in this approach in QCD stems from H.D. Politzer, Phys. Lett. 70B, 430 (1977).

42. I.G. Halliday, Nucl. Phys. B103, 343 (1976); H.D. Politzer, Nucl. Phys. B129, 301 (1977); and C.T. Sachrajda, Phys. Lett. 73B, 185 (1978) made some analyses of this process in perturbation theory. Some of their results were anticipated phenomenologically by J.B. Kogut, Phys. Lett. 65B, 377 (1976); I. Hichcliffe and C.H. Llewellyn Smith, Phys. Lett. 66B, 281 (1977). The most complete analysis is that due to Dokshitzer, D'yakonov and Troyan, ref. 10.

43. An early phenomenological discussion using this philosophy was E.G. Floratos, Nuovo Cimento 43A, 241 (1978). K.H. Craig and

C.H. Llewellyn Smith, Phys. Lett. 72B, 349 (1978) initiate a perturbative justification of the QCD calculation of large p_T jets in deep inelastic leptoproduction. Other phenomenological references are H. Georgi and H.D. Politzer, Phys. Rev. Lett. 40, 4 (1978), G. Altarelli and G. Martinelli, Rome University preprint "Transverse Momentum of Jets in Electroproduction from Quantum Chromodynamics" (1978).

44. C.L. Basham, L.S. Brown, S.D. Ellis and S.T. Love, Washington University preprint RLO-1388-746 (1977).
45. T. Kinoshita, J. Math. Phys. (N.Y.) 3, 650 (1962); T.D. Lee and M. Nauenberg, Phys. Rev. 133, B1549 (1964).
46. The correct specification of the infrared cutoff requires more care than that taken here. Consult ref. 45 and Sterman and Weinberg, ref. 9, for more details.
47. See Sterman and Weinberg, ref. 9.
48. G. Sterman, Stony Brook preprints ITP-SB-77-69, 72 (1977).
49. The basic decay modes of charmonium states are discussed in ref. 3. The observability of 3-jet final states was first discussed in ref. 38: see also references 39 and 40. More consideration should be given to the infrared singularities associated with the initial bound state before these results can be regarded as "rigorous".
50. H. Georgi and M. Machacek, Phys. Rev. Lett. 39, 1237 (1977); E. Farhi, Phys. Rev. Lett. 39, 1587 (1977).
51. Once multi-jet events are observed, one would like to study 3-jet Dalitz plots and so on (see G. Parisi, Phys. Lett. 74B, (1978)): the only problem is how to pick such events out experimentally. For a systematic analysis, see ref. 40.
52. For the moment, this and subsequent factorization statements should be taken as applying in a leading logarithmic approximation.
53. See also Dokshitzer, D'yakonov and Troyan (ref. 10) who discuss two final state hadrons.
54. There will be contributions to $\langle p_T^2 \rangle$ arising from the primordial k_T of the colliding constituents, as well as from QCD perturbative effects. Note that in QCD perturbation theory the p_T of the lepton pair is not necessarily the sum of the k_T of the colliding constituents. See E.L. Berger, Argonne preprint ANL-HEP-PR 78/12 (1978) for a recent phenomenological review.
55. H. Georgi, Harvard University preprint HUTP-77/A090 (1977) makes this warning very strongly. It will be important to do a complete phenomenological analysis. The leading logarithms in $g+q \to \ell^+\ell^- + q$ actually contribute to the $q+q \to \ell^+\ell^-$ subprocess: see Sachrajda, ref. 42.
56. C.T. Sachrajda, CERN preprint TH 2459 (1978); W. Furmanski, private communication and Krakow University preprint to appear. For a related discussion within the CIM framework, see W.E. Caswell, R.R. Horgan and S.J. Brodsky, SLAC preprint PUB-2106 (1978).
57. D. Amati, R. Petronzio and G. Veneziano, CERN preprint TH 2470 (1978) have an elegant demonstration of factorization at the one-loop level, and related arguments for both leading and non-leading

logarithms is being prepared by R.K. Ellis, H. Georgi, M. Machacek, H.D. Politzer, G.G. Ross, private communication.

58. K.M. Bitar and S.-J. Chang, Phys. Rev. D17, 486 (1978) point out that one can formulate tunnelling directly for Minkowski field theory.

59. G. 't Hooft, Phys. Rev. D14, 3432 (1976). The Pauli-Villars regularization scheme is used when evaluating the coefficients in equation (26) (see also references 13 and 14). The parameters Λ and m_i are those defined perturbatively in ref. 16: asymptotically $m_i \sim \bar{m}_i \, (\ln Q^2)^{\frac{-12}{33-2N_f}}$.

60. L.S. Brown, R. Carlitz, D. Creamer and C. Lee, Phys. Lett. 70B, 180 and Washington University Preprint RLO-1388-735 (1977).

ARE QUARKS DIRAC PARTICLES?

H. David Politzer*
California Institute of Technology, Pasadena, California 91125

ABSTRACT

Calculations of scaling properties of hadronic semi-inclusive processes are discussed using an analysis of the infrared structure of perturbative QCD.

INTRODUCTION

There can be no doubt of the permanent significance of the Dirac equation. A modest embellishment, of growing popularity over the last fifty years, is the idea that the interactions of relativistic spin-1/2 particles can be described by a local gauge principle. A crucial problem, then, is to identify what are the fundamental particles and what are the fundamental gauge groups. Historically, the prediction of the electron magnetic moment was a triumph, while the anomalous moment of the proton was really the first indication of its composite structure.

We can ask today whether the quarks, which we know from hadron spectroscopy and the probings of leptonic currents, are indeed fundamental Dirac particles. This is another way of asking, "Is QCD, the colored-quark-gluon gauge theory of strong interactions, a valid description of hadrons?"

There are only two canonical processes for which we have extracted detailed predictions from QCD. They are inclusive leptoproduction and e^+e^- annihilation. It is imperative that we find methods of making further predictions -- to test the theory more stringently, and to provide a reliable framework and reference point for the discovery and interpretation of new phenomena (e.g., in the weak interactions).

Very tentative attempts to generate new predictions have been undertaken. Most studies have simply considered the first non-trivial corrections to Born graphs[1]. I wish to describe here an analysis, applicable to all orders, of the requisite properties of QCD perturbation theory for semi-inclusive processes. This is a collaborative effort[2], and the work is in progress. I firmly believe we have the elements of the necessary proofs, but the ε's and δ's are yet to be filled in.

THE THEOREM

The result we are aiming at is the following: the cross section, $d\sigma$, for any process of the form $a \to b + X$ where a and b are sets of hadrons, currents, and/or jets with all invariants (except

*Supported in part by the A. P. Sloan Foundation and the U.S. Department of Energy under Contract No. EY76-C-03-0068.

ISSN: 0094-243X/78/108/$1.50

for masses) large is of the form

$$d\sigma(P_i) = \int \prod_i dz_i \tilde{f}_i^{a_i}(z_i,\Delta_i)\, d\tilde{\sigma}(z_i P_i,\Delta_i) + O(1/s) \tag{1}$$

where P_i are the particle momenta; $\tilde{f}_i$ are as yet uncalculable functions measured experimentally; $d\tilde{\sigma}$ is a reduced cross section calculable in perturbation theory; Δ_i is a dimensionful parameter introduced to make the separation of $\tilde{f}$ and $d\tilde{\sigma}$ precise; and s is the common scale of the large invariants. Δ_i essentially limits the transverse momenta absorbed into $\tilde{f}_i$ and provides an infrared cut-off on $d\tilde{\sigma}$. An important net result is that the $\tilde{f}_i$ are process independent.

The significance of establishing eq. (1), at least to all orders of perturbation theory, goes beyond simply justifying the important lowest order calculations[1] which already possess that form. A uniform analysis assures self-consistency. So it is possible to avoid double counting, under counting, and plain wrong counting with confidence. (I mention this because many of the preliminary phenomenological analyses have such flaws.) One such lesson from the present analysis is that the 1/s terms are not process independent. They include final state interactions, for example. Hence a careful phenomenological fit of 1/s effects in one process is not of much significance at present because we would not know its implications for another process.

An underlying assumption, not presently amenable to proof, is that the hadron wave functions are sufficiently soft to guarantee that processes that vanish like 1/s for quark and gluon scattering will continue to do so for their bound states, hadrons. Hence we concentrate on quark-gluon-current inclusive scatterings in perturbation theory. Equation (1) is represented in Fig. 1, the appropriate discontinuity of a forward amplitude.

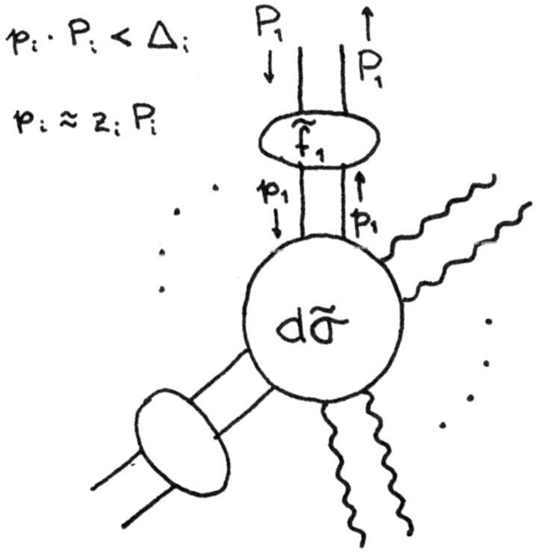

Fig. 1: The inclusive cross section for $a \to b + X$.

The obstacle to progress in the past was that in gauge theories, the infrared sensitive pieces do not organize themselves as suggested in Fig. 1. Specifically there are two-particle irreducible graphs, which apparently would be included in $d\tilde{\sigma}$, which are infrared divergent in the limit that all masses go to zero.

In fact, the graphical location and organization of infrared divergences is gauge dependent. In axial gauges ($n_\mu A^\mu = 0$ for some n_μ, preferably not $n^2 = 0$ because of $\log n^2$ terms) there is a zero associated with colinear emissions, which appears to render the two particle irreducible graphs infrared finite. The detailed arguments will be presented elsewhere[2].

IMPLICATIONS

In a qualitative way, eq. (1) is a justification of the parton picture for hard processes. The $\tilde{f}_i$ are universal distribution and decay functions that are convoluted with a parton cross section $d\tilde{\sigma}$. The quantitative differences from a most naive parton picture can be enormous, however. They arise from the following new features: Gluons, as well as quarks, must be considered as active partons. And more significantly, the computable $d\tilde{\sigma}$ now includes inelastic parton processes. These can radically alter even the most qualitative predictions, e.g., for specific phenomena that are vanishingly small in the Born approximation.

An issue that inevitably causes confusion is to what extent are radiative corrections already included in the naive parton model. A precise answer is offered by eq. (1) and Fig. 1. Indeed, $d\tilde{\sigma}$ is not the sum of all Feynman graphs. Parts of various graphs must be absorbed into the $\tilde{f}_i$, to render $d\tilde{\sigma}$ calculable (i.e., infrared insensitive). But to make the $\tilde{f}_i$ universal, process independent functions only those radiative corrections that render physically indistinguishable states (within a criterion set by Δ_i) get lumped into $\tilde{f}_i$. (From this point of view, eq. (1) is no doubt an immediate consequence of the Lee-Nauenberg theorem[3].) An example of such indistinguishability is a massless quark state compared to a quark and gluon state, both moving in the same direction. Such states can be identical in all quantum numbers. But once the quark-gluon system has an opening angle, it has a non-zero invariant mass. The lesson of eq. (1) is that the effects of radiation at finite angles is not yet included in the naive parton model but must be added into $d\tilde{\sigma}$.

Phenomenological arguments based purely on leading log calculations may be misleading because the kinematic regions in which the leading logs are genuinely large get reinterpreted and reabsorbed somewhere else. The relative sizes of the parts left over may be subtle. For example, a calculation of all Feynman graphs to a given order may yield an expression like

$$ag^4\log^2\frac{Q^2}{p^2} + bg^4\log\frac{Q^2}{p^2} + cg^2\log\frac{Q^2}{p^2} + dg^2 + eg^4$$

where $Q^2 >>> p^2 \approx 0$. But the corresponding contribution to $d\tilde{\sigma}$ would look very similar but with p^2 replaced by Δ. However, $\log Q^2/\Delta$ will typically not be enormous. Hence the "leading log" terms are not obviously dominant.

Experimental consequences, in as much as they differ from the naive parton model, have been discussed at length elsewhere[1,4]. I wish only to mention the salient features. There are logarithmic scaling violations implicit in $d\tilde{\sigma}$ of the type encountered in electroproduction. To make them explicit, note that the product $f_i(z_i,\Delta_i)d\tilde{\sigma}(s,y_j,\Delta_i)$ is independent of Δ_i, where s is a scale and y_j are dimensionless ratios of large invariants. If we take $\Delta_i = \varepsilon_i s$ with ε_i fixed, the "logarithmic" scaling violations have now moved to the $\tilde{f}_i(z_i,\varepsilon_i s)$ which depend on s in the way familiar from electroproduction.

Inelastic parton processes are of particular importance for weak and electromagnetic effects in p - p collisions because of the relative abundance of quarks and gluons compared to antiquarks. Thus processes like quark + gluon → quark + W boson (or quark + $\mu^+\mu^-$) will be competitive in pp (as well as quark + quark → W + X). As a consequence $\vec{p}_\perp^{\,2}$'s are expected to rise linearly with s.

Finally, the predictions for hadrons at large $p_\perp$, e.g., $p + p \to \pi + X$, are radically different from any previous expectations. The predictions lie dramatically above the linear fit to existing data, i.e., $p_\perp^{-8}$; and they lie dramatically below the $p_\perp^{-4}$ expected on the basis of one gluon exchange. The latter difference is due to a piling up of many logarithmic effects from next order corrections. These "go away" asymptotically but are expected to give an effective $p_\perp^{-6.\pm\varepsilon}$ for the region $6 \leqslant p_\perp \leqslant 30$ GeV. This abundance of hadrons at large $p_\perp$ (relative to the phenomenological $p_\perp^{-8}$) if observed will provide a horrendous background to weak and electromagnetic physics at high energies.

But such an observation will also provide striking evidence that the quarks are indeed Dirac particles. I would regard this as an honor for the quarks.

REFERENCES

1. H. D. Politzer, Nucl. Phys. B129, 301 (1977); Phys. Lett. 70B, 430 (1977);
   C. T. Sachrajda, CERN preprints TH.2416, 2459 (1977-78);
   G. Sterman and S. Weinberg, Phys. Rev. Letts. 39, 1436 (1977).

2. K. Ellis, H. Georgi, M. Machacek, H. D. Politzer and G. Ross, to be published.

3. T. D. Lee and M. Nauenberg, Phys. Rev. 133, 1549 (1964).

4. There are already too many to list accurately.

THE STATIC POTENTIAL ENERGY OF A HEAVY QUARK AND ANTI-QUARK*

K. Johnson
Massachusetts Institute of Technology
Cambridge, MA 02139

We have heard quite a bit about "linear" quark confining potentials in the past few years. This has been stimulated by the great phenomenological success[1] for charmonium spectroscopy obtained using a model based upon heavy quarks which move with such a potential energy. It is believed by many that the likely place to find a potential of this type is in the currently most fashionable field theory, quantum chromodynamics. This is because a potential which is proportional to the distance between field sources is familiar in ordinary electrodynamics when the electric field lines which connect the sources are lined up parallel to each other, as for example, when the sources are opposite charges located on parallel conducting plates.

Of course, nobody thinks that the quark and antiquark colors in charmonium are located on parallel plates, the belief (or hope) is that non-linear effects associated with the large color charge in QCD will lead to a focusing of the color fields into a tube of parallel "chromoelectric" flux. Now, QCD is a complicated theory so no one actually knows whether or not this happens. One might ask if there is a local way of describing an effect which will cause the color electric field to become focused short of simply hoping that some complicated non-linear consequences of the QCD interaction achieve it. Suppose there is a region of space which contains a chromo-static field with energy density $\frac{1}{2}E^2$. The Maxwell stress corresponds to a tension per unit area (↔) along the lines of flux equal to $\frac{1}{2}E^2$, and the same pressure (↕) between the lines. If the field falls to zero at some surface, as it must do if the lines are in a flux tube, then at the surface the outward pressure of the lines, $\frac{1}{2}E^2$, must be balanced by whatever effect causes them to focus. Let us <u>assume</u> that color fields can only be supported in the vacuum when they have a strength which exceeds a critical value, i.e.

$$\tfrac{1}{2}E^2 \geq B$$

where B measures the critical strength. Since this refers to a property of the vacuum, it must be a Lorentz invariant condition, i.e.

*Work supported in part through funds provided by the U.S. DEPARTMENT OF ENERGY(DOE), contract EY-76-C-02-3069.

ISSN: 0094-243X/78/112/$1.50

$$-\frac{1}{4}F^{\mu\nu}F_{\mu\nu} \geq B \qquad (1)$$

(or for non-Abelian fields $-\frac{1}{4}[F_a^{\mu\nu}F_{\mu\nu}^a] \geq B$.) In this case, at the edge of the flux tube, the pressure in the field lines will be equal to B, and hence to balance the Maxwell stress there must be a counter pressure equal to B. Since this is a property of the vacuum it will correspond to a stress tensor in the region <u>exterior</u> to the color fields equal to

$$g^{\mu\nu}B$$

where $g^{\mu\nu}$ is the Lorentz metric. This is equivalent to a stress associated with the interior region equal to $-g^{\mu\nu}B$, since a constant stress through space can have no observable consequence. The presence of such a stress is the basic postulate of the M.I.T. Bag Model.[2] We call B the "bag constant". Now we can see that a linear flux tube containing a constant field, E, can be in equilibrium. For with the addition of $-g^{\mu\nu}B$, there is a net tension per unit area along the field lines equal to $\frac{1}{2}E^2+B=2B$, (to be balanced at the sources of the lines) and a pressure between the lines equal to $+\frac{1}{2}E^2-B=0$. There is an energy density, $\frac{1}{2}E^2+B=2B$, equal to the tension per unit area. If the flux tube spans a cross section A, the net tension, T, equals 2BA. By Gauss's law, $EA=e=A\sqrt{2B}$, so $T=e\sqrt{2B}$. e measures the magnitude of the color charge on which the chromo-static flux begins and terminates. In the case of non-Abelian QCD when, for example, the flux tube connects a color 3 to $\overline{3}$, $e^2=4/3\ e_s^2$. Here e_s is the rationalized unit of charge, so $e_s^2/4\pi=\alpha_s$ is the strong coupling constant.[3] Thus, the tension is $T=\sqrt{\alpha_s}\ 4\sqrt{2\pi/3}\sqrt{B}$. An ordinary meson corresponds to a relativistically spinning flux tube,[4] and the slope parameter of the Regge mass spectrum α' is related to T by $1/\alpha'=2\pi T=8\pi\sqrt{2\pi/3}\sqrt{\alpha_s}\sqrt{B}=36.372\sqrt{\alpha_s}\sqrt{B}$. The observed slope α' is .9 GeV^{-2} so we find $\sqrt{\alpha_s}\sqrt{B} \simeq [.175\ \mathrm{GeV}]^2$. $B^{\frac{1}{4}}$ which has the dimensions of mass sets the scale of masses for ordinary spectroscopy of the hadrons and was determined to be $B^{\frac{1}{4}}\sim.145$ GeV,[5] so we find $\alpha_s\sim 2.1$. That is, the strong interaction coupling parameter is about 100 times larger than the electromagnetic parameter.

Now the flux tube in low mass mesons terminates on light quarks which are kept separated by an angular momentum barrier. Imagine instead that it connects heavy, slowly moving quarks and acts as a static potential.[6] This potential will be proportional to the distance between the quarks, with a slope equal to $T=1/2\pi\alpha'=.18\ \mathrm{GeV}^2$, which is just about equal to the slope parameter determined from charmonium phenomenology.[1] However, we would seem to be unjustified to simply assume that the flux tube forms so quickly as the quark and antiquark begin to separate. We might only expect a potential of that sort asymptotically as the distance between the quarks gets large. The relevant length scale should be the radius of the flux tube, i.e. $R=\sqrt{A/\pi}=(e/\sqrt{2B}\ 1/\pi)^{\frac{1}{2}}=\sqrt{\alpha'}\sqrt{\alpha_s}(32\pi/3)^{\frac{1}{2}}\sim 8\ \text{GeV}^{-1}$, that is, about 1.5 fermi. However, in the low lying states of the charmonium system, the quarks are separated at a distance of less than a fermi.

We would like to show that the inequality (1) leads to the effective interaction developing the linear part as soon as the charges begin to separate.[6] For convenience, we shall carry out the discussion in the language of an Abelian analogue to quantum chromodynamics. The results are equivalent to the color theory with the replacement $e^2=e_s^2\,\frac{4}{3}$, since we will only require the lowest order of QCD perturbation theory to obtain our effective potential.

Then the analogue fields[7] can be obtained from the variational principle $\delta W=0$, with

$$W = \int (d^3x)(\tfrac{1}{2}E^2-B)\,\theta(\tfrac{1}{2}E^2-B) \tag{2}$$

subject to the constraint

$$\nabla\cdot E = e(\delta^{(3)}(\vec{x} - \tfrac{\vec{r}}{2}) - \delta^{(3)}(\vec{x} + \tfrac{\vec{r}}{2})) \tag{3}$$

so that the quarks, located at the positions $\pm\frac{\vec{r}}{2}$, are the sources of the "analogue" color electric field. We have inserted into (2), $\theta(x)=\begin{cases}1 & x\geq 0\\ 0 & x<0\end{cases}$ to enforce the constraint $\frac{1}{2}E^2>B$. For consistency, we have added the term $-B$ to $\frac{1}{2}E^2$ to make the integrand a continuous function of the field strength. We shall see that this simply corresponds to the addition of a uniform pressure exterior to the domain containing the fields. We have noted already the necessity of this term. If we require that (2) be stationary on arbitrary variation of E, we find,

$$\vec{E}\theta = -\nabla\phi \tag{4}$$

where ϕ is a Lagrange multiplier introduced to maintain the constraint (3) on $\vec{\nabla}\cdot E$. Then from (3) we obtain

$$-\nabla^2\phi=\vec{\nabla}\cdot(\vec{E}\theta)=\vec{\nabla}\cdot\vec{E}+\vec{E}\cdot\nabla(\theta)=e(\delta^{(3)}(\vec{x}-\tfrac{\vec{r}}{2})-\delta^{(3)}(\vec{x}+\tfrac{\vec{r}}{2}))+\hat{n}\cdot\vec{E}\delta(\tfrac{1}{2}E^2-B)\lambda \tag{5}$$

where $\hat{n}$ is normal to the surface whose equation is $\frac{1}{2}E^2-B=0$, and beyond which the fields must vanish. That is $\nabla(\frac{1}{2}E^2-B)=\hat{n}\lambda$ on the surface. We see ϕ, where

$$-\nabla^2\phi = e(\delta^{(3)}(\vec{x}-\frac{\vec{r}}{2})-\delta^{(3)}(\vec{x}+\frac{\vec{r}}{2})) \qquad (6)$$

with $\hat{n}\cdot\nabla\phi=0$ on the surface, gives a unique solution to the problem. As already noted,

$$\tfrac{1}{2}E^2 = \tfrac{1}{2}(\nabla\phi)^2 = B \qquad (7)$$

is the equation for the surface which encloses the domain where the fields are non-vanishing. (This is the "bag" of the M.I.T. Bag Model.) We see that (7) corresponds to locally balancing the pressure of the static field by the bag on the surface confining pressure B. It is also possible to show with a little computation that the surface which encloses the fields and on which the potential obeys (7) can be gotten from an energy variational principle. Consider,

$$U = \int_{\text{Bag}} d^3x\ (\tfrac{1}{2}(\nabla\phi)^2+B) \qquad (8)$$

where

$$-\nabla^2\phi = e(\delta^{(3)}(\vec{x}-\frac{\vec{r}}{2})-\delta^{(3)}(\vec{x}+\frac{\vec{r}}{2})) \qquad (9)$$

and where

$$\hat{n}\cdot\nabla\phi = 0 \qquad (10)$$

on the surface which encloses the "bag", the region of space included in the integral (8). Since there is a unique ϕ associated with each surface, (8) may be regarded as a surface functional. On the surface which minimizes (8),

$$\tfrac{1}{2}(\nabla\phi)^2 = B,$$

that is, we have the solution for the boundary condition (7). If we evaluate U for this case we obtain the desired static potential, $U_o(r)$. We shall show that $U_o(r)$ is very accurately (but not exactly) represented by the function

$$U_o(r) = -\frac{e^2}{4\pi r} + r\sqrt{2Be^2}\ . \qquad (11)$$

That is, the effect of imposing the constraint

$$\tfrac{1}{2}E^2 \geq B$$

is simply to add to the Coulomb chromo-static energy a linear function of r. We have already indicated how this form obtains in the limit $r \to \infty$, when the flux arranges itself in a tube. But we shall now see that (11) works over the entire range.

Suppose the surface which minimizes (8) is given by the equation

$$|\vec{x}| = R(\hat{x};r) \tag{12}$$

when the quarks are separated by the distance r. Further let the field potential be given by the function

$$\phi_o(\vec{x};r), \tag{13}$$

so

$$-\nabla^2 \phi_o = e(\delta^{(3)}(\vec{x}-\tfrac{\vec{r}}{2}) - \delta^{(3)}(\vec{x}+\tfrac{\vec{r}}{2}))$$

and $\hat{n}\cdot\nabla\phi_o = 0$ on the surface (12). Now consider the scaled surface,

$$|\vec{x}| = \frac{1}{\lambda} R(\hat{x};\lambda r). \tag{14}$$

If we use (14), to calculate (8) for any $\lambda \neq 1$ it will give a larger energy than U_o. It is easy to check that the potential appropriate to (14) is,

$$\phi_\lambda(x;r) = \lambda\phi_o(\lambda\vec{x};\lambda r).$$

That is, ϕ_λ obeys (9), and (10) on the surface (14). If we evaluate (8) using the scaled surface, we find

$$U_\lambda = \lambda S(\lambda r) + \frac{1}{\lambda^3} BV(\lambda r)$$

where

$$S(r) = \int_{B_o} d^3\vec{x}\ \tfrac{1}{2}(\nabla\phi_o)^2$$

and

$$V(r) = \int_{B_o} d^3\vec{x}.$$

B_o is the region of space enclosed by the surface (12). That is, S and BV are the field and volume energies corresponding to the minimum energy surface. Since $\partial U/\partial\lambda = 0$ at $\lambda = 1$, we find

$$0 = \frac{\partial}{\partial r}(r[S(r)+BV(r)]) - 4BV(r) \tag{15}$$

and thus by integration of (15)

$$U_o(r)=S(r)+BV(r)=-e^2/4\pi r+4B\frac{1}{r}\int_o^r dr'V(r') \qquad (16)$$

The constant of integration $-e^2/4\pi$ is obtained from the requirement that the interaction be Coulombic as $r\to 0$. (The divergent self energies have been subtracted from S.)

The second term in (16) represents the consequence of the condition $\frac{1}{2}E^2>B$. As $r\to\infty$, we have already noted that $V(r)\to Ar=e/\sqrt{2B}\ r$, and (16) of course is consistent with the linear flux tube since $U_o(r)\to e\sqrt{2B}\ r$. However, V(r) is also linear as $r\to 0$. We can see why intuitively. As $r\to 0$ we expect that the field is approximately of the dipole form, so on the surface

$$E \simeq \frac{re}{R^3 4\pi}$$

where R is the distance to the surface. By (7), on the surface

$$\sqrt{2B} = E \simeq \frac{r}{R^3}\ e\ \frac{1}{4\pi}\ ,$$

and so we find the volume $\sim 4\pi/3\ R^3$ is of order $r\ e/\sqrt{2B}\ \frac{1}{3}$. Thus, V(r) is also a linear function near r=0, with perhaps a different numerical coefficient than that given by our guess of the dipole field. We have computed $U(r) + e^2/4\pi r$ for a sphere and a cylinder as trial surfaces. In both cases we use the exactly calculable potential which is appropriate. We obtain the best sphere by minimizing (7) with respect to the radius, and the best cylinder by minimizing with respect to length and diameter. For the sphere $U(r)+e^2/4\pi r\to\sqrt{2/3}\sqrt{2B}e^2 r$ as $r\to 0$, ($\sqrt{2/3}$=.816) and for the cylinder, $\to(.840)\sqrt{2B}e^2 r$. Thus we anticipate that V(r) in (16) is of the form $V(r)=e/\sqrt{2B}\ rk(r)$ where as $r\to\infty$, $k(r)\to 1$ and where as $r\to 0$, $k(r)\sim .8$. Thus the volume energy is almost exactly linear and hence we can understand the accuracy of the form (8). These simple considerations are supported by an exact calculation which can be done for the equivalent problem in two space dimensions,[8] and by detailed computer calculations in three space dimensions.[6] We may remark that a sphere doesn't look very much like a linear tube of flux. In the case of the cylindrical approximation near r=0, the cylinder is shorter than it is thick ($L/D\sim .74$ near r=0).

Now what should we make of this? We have found in the case of heavy quarks where a static potential makes sense, that its form will be quite accurately linear plus Coulombic if the particles interact through a gauge field which obeys the restriction $\frac{1}{2}E^2 \geq B$. Thus, B determines a critical field strength. What has this got to do with Dirac? The M.I.T. Bag Model is closely related to a model studied some time ago by Dirac.[9] The term in the Lagrangian which makes the fields "lump" into a finite volume around the static charges corresponds to a term in the action equal to

$$-B\int_{Bag} d^4x \tag{17}$$

where the integral (17) extends over the tube in space-time which is swept out by the volume occupied by the fields belonging to the extended particle. The action (17) is the three dimensional version of an action for an extended particle considered some time ago by Dirac, who associated a geometrical action with the surface of an extended particle. In Dirac's case, the geometrical term has the form,

$$-\sigma\int dt\int d^2S\sqrt{1-V_T^2} \tag{18}$$

where V_T is the component of velocity of a point on the surface along the space normal to the surface element d^2S. The factor $\sqrt{1-V_T^2}$ is needed to make the above action relativistically invariant. Thus, by introducing a term into the action associated with a geometrical extension for the particle, in the case of "surface tension" one must make the geometrical variables dynamical. However, in (18) there is kinetic energy associated only with motion transverse to the extension. Such geometrical actions have also played a role in the dual string model, where the geometrical extension is one dimensional.[10]

In the M.I.T. model for hadrons the geometrical action, by itself, cannot have any dynamics associated with it since there is no kinetic energy which belongs to the internal spatial points. We made the model physical by fixing on these points field operators for creating and destroying elementary Dirac particles, in this case, quarks. Thus, hadrons, the particles which have this structure, are extended blobs carrying quark fields. As we originally conceived our model[2] we assumed that hadrons were extended "particles" carrying the geometrical term (17) in their action, along with Dirac fields attached to the geometry. However, as we have seen here, an alternative formulation can be achieved within the context of ordinary field theory, if we assume[11] that the color gauge fields obey the restriction (1). In this case the fields must necessarily lump around their sources which are then forced to be in colorless combinations.

REFERENCES

1. E. Eichten, K. Gottfried, K. Lane, T. Kinoshita, T.M. Yan, Cornell preprint #375, to appear in Phys. Rev. D.
2. A. Chodos, R.L. Jaffe, K. Johnson, C.B. Thorn, V.F. Weisskopf, Phys. Rev. D9, 3471 (1974).
3. We have adopted the definition of the coupling constant favored by the majority, α_s. It is related to that used in our earlier work, α_c, by $\alpha_c = \alpha_s/4$, or $e_c = \frac{1}{2} e_s$.
4. K. Johnson and C.B. Thorn, Phys. Rev. D13, 1934 (1976).
5. T. DeGrand, R.L. Jaffe, K. Johnson, J. Kiskis, Phys. Rev. D12, 2060 (1975).
6. P. Gnädig, P. Hasenfratz, J. Kuti, A.S. Szalay, Phys. Letters, 64B, 62 (1976).
7. These considerations will be treated in more detail in a subsequent publication.
8. R. Giles, M.I.T. preprint, CTP #706 (1978).
9. P.A.M. Dirac, Proc. R. Soc. A268, 57 (1962).
10. Y. Nambu, Lectures at the Copenhagen Summer Symposium (1970). (Unpublished).
11. K. Johnson, M.I.T. preprint, CTP #719 (1978).

NON-INTERNAL GAUGE THEORIES

Yuval Ne'eman
Department of Physics and Astronomy, Tel-Aviv University *
Tel-Aviv, Israel

and

Center for Particle Theory, University of Texas
Austin, Texas 78712, U.S.A.

ABSTRACT

We discuss Gravity and Supergravity as gauge theories constructed on their relevant Group Manifolds, in a formalism recently introduced by T. Regge and the author. We emphasize the role of Parallel-transport gauges (AGCT) in these theories and present the alternatives. We comment on possible extensions, emphasizing the Affine and Graded Affine theories, and alternatives to Einstein's Lagrangian.

INTRODUCTION

Gauge theories were originally conceived as extensions of Einstein's idea of General Covariance and of his General Theory of Relativity [1]. However, taking the viewpoint of XXth Century Differential Geometry (rather than that of Gauss and Riemann), General Covariance becomes somewhat trivial. Geometrical information is fed into systems anholonomically, using Lie groups acting on local frames, or more generally Fiber Bundles. Gauging an internal symmetry group G is just such an operation, involving a Principal Bundle $\{B, M^4, G\}$ with base space $M^4 := R^{3.1}$ (Minkowski space) and structure group G. The gauge transformations are the $g_{\alpha\beta}(x)$, $x \in U_\alpha \cap U_\beta$ (the U_α are coordinate patches), $g \in G$, so that we have a continuous map $g_{\beta\alpha}: U_\alpha \cap U_\beta \to G$. They rotate the fiber (here G itself) on the left and act on B on the right, with trivial action on M^4. B represents the sourceless case, and the $\psi(x)$ matter fields reside as sections in an Associated Vector Bundle $\{E, B, M^4, \psi, G\}$ where ψ^a is a Unitary representation of G.

Ushering in the "XXth Century" viewpoint, Cartan reformulated Einstein's theory, using the Cotangent Bundle of M^4 and in fact gauging the Lorentz group. Thirty years later, following the Yang-Mills paper, Utiyama, Sciama and Kibble attempted to derive Gravitation as a Gauge Theory. They ended up perfecting the Lorentz gauge and reproducing the Einstein-Cartan theory[2]. In fact, their gauge group was $P/T^{3.1}$, where P is the Poincaré group and $T^{3.1}$ the translations, also an M^4. They were thus gauging Lorentz spin L, since the essence of a gauge transformation is in its leaving $x \in M^4$ fixed [3].

* Research supported in part by the United-States - Israel Binational Foundation, (B.S.F.) Jerusalem, Israel.

† Research supported in part by (U.S.) N.S.F. Grant 40768X

ISSN: 0094-243X/78/120/$1.50

This emphasis on a Gauge group rather than on General Covariance is also the more physical approach: General Coordinate Transformations are just a relabelling, i.e. passive transformations. On the other hand, Gauge transformations have a limit where they loose their x dependence and correspond to a "trivial section", i.e. a global transformation. This is the symmetry which corresponds to Noether's theorems and has both active and passive realizations. Note however that it is only this global limit of the Gauge Symmetry that has a physical active realization. Gauge transformations relate field variations at points with a space-like separation, i.e. where information cannot propagate. However, they can also be regarded as relating the field (a section) in two different coordinate systems (the U_α patches), for the same point by a $g_{\alpha\beta}(x) \in G$. More recently, the theory of Supergravity [4] has been reproduced in Superspace $R^{3.1/2.2}$, with only Lorentz-spin L as a local gauge [5], i.e. $GP/_{T^{3.1/2.2}}$ where GP is the Graded-Poincaré group, and $T^{3.1/2.2}$ its ideal, constructed from the translations and supersymmetry transformations. Another type of (Parity non conserving) Supergravity can be constructed with a gauge group $GP/_{T^{3.1/2R}}$, where $T^{3.1/2R}$ is an ideal containing translations plus two supersymmetry generators, both right-chiral. Aside from Lorentz spin the gauge group L_L then contains the anholonomic action of the two left-handed supersymmetry generators.

Looking at all these examples, it becomes obvious that the relevant manifold is the Group Space. In Gravity, the dimensionality of the relevant B is $d(M^4) + d(L) = 4 + 6 = 10$. In Supergravity, it is either $d(R^{3.1/2.2}) + d(L) = 8 + 6 = 14$, or $d(R^{3.1/2R}) + d(L_L) = 6+8=14$. In fact, the base spaces M^4, $M^{4/4} = R^{3.1/2.2}$, or $R^{3.1/2R}$ are just the parameter-spaces of $T^{3.1}$, $T^{3.1/2.2}$ or $T^{3.1/2R}$.

It is as a result of this intuitive picture that with T. Regge we have recently developed [6] a formalism for Non-internal Gauge Groups (groups having some action on M^4). The arena is the Group Manifold, which automatically always contains space-time. However, the Action is integrated over M^4 only.

FORMS

The idea is to write such a theory on a group manifold. Let G be [a non semi-simple] Lie group of dimension n [e.g. the Poincaré group P for gravity, the Graded Poincaré [7] formal-group [8] GP for supergravity]. On G we give a set of n forms ρ^A and define G-curvature as the 2-forms

$$R^A = d\rho^A - \tfrac{1}{2}\, \rho^B \wedge \rho^E \, C^A{}_{\cdot BE} \qquad (1)$$

where $C_{BE}^{\cdot\cdot\; A} = -(-1)^{be}\, C_{EB}^{\cdot\cdot\; A}$ are the (graded, with b,e the gradings [9] of B,E) structure constants of the Lie algebra $\mathfrak{g}$ of G. For the case $\rho^A = \omega^A$, where ω^A are the left invariant Cartan forms on G, $R^A = 0$ and (1) realizes the Cartan-Maurer equations. We refer to this case as the "flat" geometry.

We regard ρ^A as the "objects" of the theory, to be treated as the potentials of a Yang-Mills G-gauge theory, and the R^A are the corresponding field-strengths. Given any G-multiplet η^A (η_A) we define G-covariant derivatives as,

$$(D\eta)^A = d\eta^A - \rho^B \wedge \eta^E \, C^A{}_{\cdot BE}$$
$$(D\eta)_A = d\eta_A + \rho^B \wedge \eta_E \, C^E{}_{\cdot BA} \qquad (2)$$

In the actual cases of Gravity and Supergravity, we have ({ } represent terms to be added to the Gravity case when turning to Supergravity)

$$R^{cd} = d\rho^{cd} + \rho^{ca} \wedge \rho^{ad}$$
$$R^c = d\rho^c + \rho^{ca} \wedge \rho^a \quad \{+ \tfrac{1}{2} \bar{\rho}\, \gamma^c \wedge \rho\} \qquad (3)$$
$$\{R = d\rho + \tfrac{1}{2} \rho^{ab} \, \sigma^{ab} \wedge \rho\}$$

$$(D\eta)^{ab} = d\eta^{ab} + \rho^{at} \wedge \eta^{tb} - \rho^{tb} \wedge \eta^{at}$$
$$(D\eta)^a = d\eta^a + \rho^{at} \wedge \eta^t - \rho^t \wedge \eta^{at} \quad \{+ \bar{\rho}\, \gamma^a \, \eta\} \qquad (4)$$
$$\{(D\eta) = d\eta + \tfrac{1}{2}(\rho^{ab} \, \sigma^{ab}) \wedge \eta - \tfrac{1}{2} \sigma^{ab} \, \rho \wedge \eta^{ab}\}$$

$$(D\eta)_{ab} = d\eta_{ab} - \rho^{ta} \wedge \eta_{tb} + \rho^{tb} \wedge \eta_{ta} + \rho^b \wedge \eta_a \quad + \{\tfrac{1}{2}\tilde{\rho}\, (\sigma^{ab})^{\sim} \, C\eta\}$$
$$(D\eta)_a = d\eta_a - \rho^{ta} \wedge \eta_t \qquad (5)$$
$$\{(D\bar{\eta}) = d\bar{\eta} - \tfrac{1}{2} \rho^{ab} \, \bar{\eta}\, \sigma^{ab} + \bar{\rho}\, \gamma^a \, \eta_a\}$$

In the "flat" limit,

$$\omega^{ab} = \Xi^{ca} \, d\Xi^{cb}$$
$$\omega^a = \Xi^{ca} \, dx^c - \{\tfrac{1}{2} \, \Xi^{ca} \, \bar{\xi}\, \gamma^c \, d\xi\} \qquad (6)$$
$$\{\omega = u^{-1}(\Xi) \, d\, \xi\}$$

where elements of P(GP) are given by

$$P : (Z,z) = (\Xi\Theta, \Xi t + x) = (\Xi,x)\,(\Theta, t)$$
$$GP: (Z,z,\zeta) = (\Xi\Theta, \Xi t + x + \tfrac{1}{2}\, \bar{\xi} \gamma u(\Xi)\nu,\ u(\Xi)\nu + \xi) = (\Xi,x,\xi)\,(\Theta,t,\nu) \qquad (7)$$

and superspace $R^{3.1/2.2}$ is the set of equivalence classes of GP under Right-multiplication by the Lorentz Group elements.

We can also write the L.I. vector fields (inducing right-translations), orthonormal to the ω^A,

$$D_{ab} = \Xi^{ca} \frac{\partial}{\partial \Xi^{cb}} - \Xi^{cb} \frac{\partial}{\partial \Xi^{ca}}$$

$$D_a = \Xi^{ca} \frac{\partial}{\partial x^c} \tag{8}$$

$$\{D = u^{-1}(\Xi) \left(\frac{\partial}{\partial \bar{\xi}} - \frac{1}{2} \frac{\partial}{\partial x^m} \gamma^m \xi\right)\}$$

It is also instructive to write the R.I. vector fields (inducing left-translations),

$$S_{ab} = x^a \frac{\partial}{\partial x^b} - x^b \frac{\partial}{\partial x^a} + \Xi^{af} \frac{}{\partial \Xi^{bf}} - \Xi^{bf} \frac{}{\partial \Xi^{af}} + [(\bar{\xi}\, \sigma^{ab} \frac{\partial}{\partial \bar{\xi}})]$$

$$S_a = \frac{\partial}{\partial x^a} \tag{9}$$

$$\{S = \frac{\partial}{\partial \bar{\xi}} + \frac{1}{2} \frac{\partial}{\partial x^m} \gamma^m \xi\}$$

Note that the R.I. left-translations vector fields correspond to our conventional generators acting on fields etc., with S_{ab} as total angular momentum, whereas the L.I. set has only spin in D_{ab}. These L.I. generators act on the right, and due to the definition of multiplication in P and GP given in eq. (7), we can see that Θ only acts on Ξ; indeed, taking $\Xi = 0$, so as to observe the action-on the right on a space-time coordinate $(0,x)$ we see that $(0,x) \to (0,x)$ and the D_{ab} indeed fit the role of a gauge-group in a Fiber Bundle. On the other hand, taking $\Theta = 0$ and identifying (o,t) with a space-time coordinate shows that the left-acting S_{ab} indeed rotates the coordinate $t \to \Xi t$, i.e. it contains orbital angular momentum.

GAUGE TRANSFORMATIONS AND FACTORIZATION

Given manifolds M and N, a map λ: $M \overset{\lambda}{\to} N$, and a form ν^P on N,

$$\nu = \sum_{(A)} dy^{A_1} \wedge \dots \wedge dy^{A_p} N_{A_1 \dots A_p}(y), \quad y \in N \tag{10a}$$

the pullback $\lambda^* \nu^P$ on M is given (for $y = \lambda(x)$) by

$$\lambda^*\nu = \sum_{(B)} dx^{B_1} \frac{\partial y^{A_1}}{\partial x^{B_1}} \wedge \ldots \wedge dx^{B_p} \frac{\partial y^{A_p}}{\partial x^{B_p}} N_{(A)} (\lambda(x)) \tag{10.b}$$

Using the group G action on either flank,

$$\begin{aligned} \ell(a)x &= ax \\ r(a)x &= xa \end{aligned} \tag{11}$$

we have for L.I. forms ω

$$\begin{aligned} \ell^*(a)\omega &= \omega \\ r^*(a)\omega &= \omega \text{ ad } (a^{-1}) \end{aligned} \tag{12}$$

where ad(a) is the adjoint representation of G. Taking the group-multiplication itself as a map Λ: G x G $\overset{\Lambda}{\to}$ G, we find

$$\Lambda^*\omega = \omega_y + \omega_x \text{ ad } (y^{-1}) \tag{13}$$

We now turn to a group multiplication in which a map is defined between the two copies of the group, : $G_x \xrightarrow{\Lambda} G_y$; or $y = \lambda(x)$ where $(x)(y) = (xy)$ is the group multiplication. This is a <u>gauge transformation</u> since we can regard the copy y as the transformation generated by the D_A right translations, and the copy x as the base manifold. On the set of "vielbeins"* ρ^A we introduced in (1), we have,

$$\Gamma(\lambda)\ \rho_x^{\ A} = \lambda^*\ \omega_y^{\ A} + \rho_x^{\ B} \text{ ad } (\lambda^{-1}(x))_B^{\ A} \tag{14a}$$

$$\Gamma(\lambda)\ R_x^{\ A} = R_x^{\ A} \text{ ad } (\lambda^{-1}(x))_B^{\ A} \tag{14b}$$

This is equivalent to a gauge in which the entire group G (i.e. P or GP) would be acting on the entire manifold G, i.e. a Bundle with 2n dimensions. However, given a Lie subgroup $H \subset G$, we can choose a set of transformations (14) such that the forms $\rho_x^{\ B}$ in (14a) will become one-forms τ^B on the homogeneous space M only,

$$x \in M = G/H$$

Using the Principal Bundle {G, M, H} (see our introduction) with π the projection $G \to M$, U_α coverings of M, $\pi^{-1}(U_\alpha) := (x, f_\alpha, \alpha)$ and transition functions $h_{\alpha\beta}(x)$, with $f_\alpha = h_{\alpha\beta} f_\beta$, we define a map χ_α projecting onto the fiber $f \in H$

* We are indebted to Prof. M. Gell-Mann for this linguistic extension of the term "vierbein" (tetrad).

$$\chi_\alpha : p^{-1}(U_\alpha) \xrightarrow{\chi_\alpha} G$$
$$\chi_\alpha\ (x,\ f_\alpha,\ \alpha) = f_\alpha \qquad (15)$$

Equations (14) can now be rewritten for that "factorized" situation,

$$\rho_\alpha{}^A = \chi_\alpha^* \, \omega^A + \tau_\alpha{}^B \, ad(f_\alpha^{-1})_B{}^A \qquad (16a)$$

with a "matching condition",

$$\tau_\alpha{}^A = h_{\beta\alpha}^* \, \omega^A + \tau_\beta{}^B \, ad(h_{\beta\alpha}^{-1})_B{}^A \qquad (16b)$$

For the factorized vielbeins τ^A of (16a), (16b) represents an H-Gauge transformation. For P and GP, taking the Lorentz group L (SO(1,3)) as H, the factorization equations (16a) yield,

$$\rho^{ab}(\Xi,x,\{\xi\}) = (\Xi^{-1}\, d\Xi)^{ab} + \tau^{cd}(x,\{\xi\})\,\Xi^{db}\,\Xi^{ca}$$
$$\rho^{a}(\Xi,x,\{\xi\}) = \Xi^{ca}\,\tau^{c}(x,\{\xi\}) \qquad (17)$$
$$\{\rho\ (\Xi,x,\xi) = u^{-1}(\Xi)\ \tau\ (x,\xi)\}$$

$$R^{ab} = \Xi^{db}\,\Xi^{ca}\,\mathcal{R}^{cd}$$
$$R^{a} = \Xi^{ca}\,\mathcal{R}^{c} \qquad (18)$$
$$\{R = u^{-1}(\Xi)\,\mathcal{R}\}$$

where the curly $\mathcal{R}^A$ are two-forms in x (or (x,ξ) for GP) whose explicit structure is identical with the expressions in (3), except that all $\rho^A(\Xi,x,\{\xi\}) \to \tau^A(x,\{\xi\})$. Infinitesimally, (14a) becomes for the entire (unfactorized) G (using (4))

$$\delta\rho^A = D\,\varepsilon^A$$

or for P{GP}, $\varepsilon^{ab} = U^{ab} - \delta^{ab}$, $\varepsilon^a = u^a$ $\{\varepsilon = u\}$

$$\delta\rho^{ab} = d\varepsilon^{ab} + \rho^{ac}\,\varepsilon^{cb} - \varepsilon^{ac}\,\rho^{cb} = D\varepsilon^{ab}$$

$$\delta\rho^{a} = d\varepsilon^{a} + \rho^{ac}\,\varepsilon^{c} - \varepsilon^{ac}\,\rho^{c} + \{\bar{\rho}\,\gamma^a\,\varepsilon\} = D\varepsilon^{a} \qquad (19)$$

$$\{\delta\rho = d\varepsilon + \tfrac{1}{2}\rho^{ab}\,\sigma^{ab}\,\varepsilon - \tfrac{1}{2}(\sigma^{ab}\rho)\,\varepsilon^{ab} = D\,\varepsilon\,\}$$

For the factorized potentials we have the Lorentz-gauge,

$$\delta\tau^{ab} = D\varepsilon^{ab} = D^{L}\,\varepsilon^{ab}$$

$$\delta\tau^{a} = -\varepsilon^{ac}\,\tau^{c} \qquad (20)$$

$$\delta\tau = -\frac{1}{2}\,\varepsilon^{ab}\,(\sigma^{ab}\tau)$$

PARALLEL TRANSPORT GAUGES (AGCT)

We now come to one of the more difficult issues: gauging the translations. Considering that the entire gravitational interaction centers on the coupling of the gravitational field ($g_{\mu\nu}$ in Einstein's holonomic - "world tensor" system) to the density of Poincaré translations (the energy-momentum tensor) it seemed disappointing that the gauge method should involve spin rather than translations.

Working in the Einstein-Cartan theory with that spin gauge still involves the cotangent bundle of space-time, with vierbein forms

$$e^{a} = e^{a}{}_{\mu}\,dx^{\mu}\,, \quad g_{\mu\nu} = \eta_{ab}\,e^{a}{}_{\mu}\,e^{b}{}_{\nu} \qquad (21)$$

$$\delta e^{a}{}_{\mu} = (\partial_{\mu}\,\varepsilon^{\nu})\,e^{a}{}_{\nu}$$

where the variation relates to a General Coordinate transformation (GCT). However, GCT as such are unrelated to the Poincaré group (or to GP) and do not represent a gauge with a global symmetry limit. Indeed, GCT contain transformations such as $x \to \lambda x$ even if the theory is not scale invariant!

Using the gauge defined in (14), or (17)

$$\delta\rho^{a} = D\varepsilon^{a} \;; \qquad \delta\tau^{a} = D^{L}\,\varepsilon^{a} \qquad (22)$$

$$\{\delta\rho = D\varepsilon\} \;; \qquad \{\delta\tau = D^{L}\,\varepsilon\}$$

In these expressions, e^{a} has been replaced by τ^{a}, a relationship known as a "Cartan connection", in which the frame forms (vierbeins) become gauge-potentials, i.e. "connections", in an Extended Affine Bundle in which $H \subset G$ has been replaced by G itself. However, it turns out that neither the Einstein Lagrangian nor any of its allowed alternatives is indeed invariant under (22). The same is true of Supergravity.

The correct translation gauge corresponds in fact to a modified Poincaré algebra, in which the commutation relations of the horizontal generators are replaced by expressions involving curvatures R^{A} as structure functions. We begin by calculating the expressions for an orthonormal set of vector fields D_{A} orthogonal to the ρ^{A} of (1), i.e.

replacing the D_A of eq. (8) which were orthogonal to the L.I. limit of ρ^A, the ω^A. The curvature (1) definition can be written as a "pseudo Cartan-Maurer equation" with $R^A{}_{\cdot BE}$ entering as "structure-functions".

$$d\rho^A - \frac{1}{2}\rho^B \wedge \rho^E (C^A{}_{\cdot BE} + R^A{}_{\cdot BE}) = 0 \tag{23}$$

The vector fields defined by

$$\rho^A (\tilde{D}_B) = \delta^A{}_{\cdot B} \tag{24}$$

have commutation relations

$$[\tilde{D}_B, \tilde{D}_E] = (C^A{}_{\cdot BE} + R^A{}_{\cdot BE}) \tilde{D}_A \tag{25}$$

so that

$$\begin{aligned} (\delta\tilde{D})_E &= [\varepsilon^B \tilde{D}_B, \tilde{D}_E] = \varepsilon^B (C^A{}_{\cdot BE} + R^A{}_{\cdot BE}) \tilde{D}_A \\ (\delta_B \tilde{D})_E &= (C^A{}_{BE} + R^A{}_{BE}) \tilde{D}_A \end{aligned} \tag{26}$$

and from (24), with b, e gradings of B, E in the GP case,

$$\delta_B\rho^A = -(-1)^{be} \rho^E (C^A{}_{\cdot EB} + R^A{}_{\cdot EB}) \tag{27}$$

This is the variation of the adjoint representation ρ^A as extracted from that of the co-adjoint $\tilde{D}_A$. Considering (22) as a generator algebra, and adding a gauge term $d\varepsilon^A$ for $\varepsilon^A(\Xi,x)$ we find

$$\begin{aligned} \delta\rho^A &= d\varepsilon^A - (-1)^{be} \rho^E \varepsilon^B (C^A{}_{\cdot EB} + R^A{}_{\cdot EB}) = \\ &= D\varepsilon^A - \rho^E \varepsilon^B R^A{}_{\cdot EB} \end{aligned} \tag{28}$$

and for the factorized case, with A, B ε [H]; E, F ε [G/H] when [ ] stands for the range of indices,

$$\rho^F (D_A) = 0$$

$$\rho^B (D_A) = \omega^B (D_A) + \tau^B (D_A) \, ad(h^{-1})_B{}^A$$

but since τ^B contains only G/H differentials, the second term vanishes, so that

$$\rho^{B}\ (D_A) = \omega^{B}\ (D_A)$$

$$\rho^{F}\ (D_A) = \omega^{F}\ (D_A)$$

$$D_A = \tilde{D}_A \tag{29a}$$

$$R^{A}{}_{BE} = 0, \qquad \forall\ c \in [G] \tag{29b}$$

Thus the Lorentz group L preserves its "flatness" in both P and GP. For the horizontal generators we find, after factorization

$$\begin{aligned}\tilde{D}_F &= \mathrm{ad}(h)_F{}^{E}\ T_E{}^{\mu}\ \left(\frac{\partial}{\partial Z^{\mu}} - \frac{1}{2}\sum_{[H]} \tau_{\mu}{}^{A}\ S_A\right)\\ &= \tilde{\Delta}_F{}^{\mu}\ D^{(H)}_{\mu}\end{aligned} \tag{30}$$

where $Z^{\mu} \in M = G/H$, i.e. it is a space-time coordinate for P or a superspace coordinate for GP;

$$\rho^{E}\ (T_F) = \tau^{E}\ (T_F) = \delta^{E}{}_{\bullet F};\ \rho^{A}\ (T_F) \neq \delta^{A}{}_{F}$$

S_A are the R.I. generators for the (Lorentz) subgroup H, and $D^{(H)}_{\mu}$ is the conventional covariant derivative.

We now analyze these horizontal transformations and gauges. Taking a GCT we can unite (still in the unfactorized situation, ζ^M a coordinate over the whole of G, and $A \in [G]$)

$$\delta\zeta^{M} = \varepsilon^{M}$$

$$\begin{aligned}\delta\rho^{A} &= \delta(d\zeta^{M}\ \rho^{A}{}_{\bullet M}) = d\zeta^{N}\ \frac{\partial\varepsilon^{M}}{\partial\zeta^{N}}\ \rho^{A}{}_{\bullet M} + d\zeta^{M}\ \varepsilon^{N}\ \frac{\partial}{\partial\zeta^{N}}\ \rho^{A}{}_{\bullet M}\\ &= d\zeta^{N}\ \left[\frac{\partial\varepsilon^{A}}{\partial\zeta^{N}} + \varepsilon^{M}\ \frac{\partial}{\partial\zeta^{M}}\ \rho^{A}{}_{\bullet N} - (-1)^{mn}\ \varepsilon^{M}\ \frac{\partial}{\partial\zeta^{N}}\ \rho^{A}{}_{\bullet M}\right]\end{aligned}$$

where we have defined our anholonomized G.C.T.

$$\varepsilon^{A} = \varepsilon^{M}\ \rho^{A}{}_{\bullet M} \tag{31}$$

We can rewrite

$$\delta\rho^{A} = d\varepsilon^{A} - 2\ (\varepsilon,\ d\rho^{A})$$

where in

$$d\rho^A = -\frac{1}{2}(d\zeta^N \wedge d\zeta^M)\left(\frac{\partial}{\partial\zeta^M}\rho^A{}_{\cdot N} - (-1)^{mn}\frac{\partial}{\partial\zeta^N}\rho^A{}_{\cdot M}\right)$$

the ε^M is contracted with the second factor. We thus get ($B,E \ \varepsilon [G]$)

$$\delta\rho^A = D\varepsilon^A + C^A{}_{\cdot BE}\,\rho^B \wedge \varepsilon^E - 2\,(\varepsilon,\ d\rho^A)$$

$$= D\varepsilon^A - 2\,(\varepsilon,\ R^A) = D\varepsilon^A - \rho^B\,\varepsilon^C\,R^A{}_{\cdot BC} \qquad (32)$$

Thus, the A.G.C.T. realizes the $\tilde{D}_F$ algebra (parallel transport generators) gauge. Formula (32) can be taken as the most general case, since for $A\ \varepsilon\ [H]$, $c\ \varepsilon\ [H]$, $B\ \varepsilon\ [G]$, equation (29b) guarantees the vanishing of the curvature term for ε^c in the factorized case. However, the horizontal ε^F will contribute such curvature terms for any $R^A{}_{.EF}$ (where $E,\ F\ \varepsilon\ [G/H]$).

Our analysis guarantees the exponentiation of the AGCT since we know that the group is in fact the GCT group.

For P and GP we have,

$$\delta\rho^{ab} = D\varepsilon^{ab} - 2\,(\varepsilon,\ R^{ab})$$

$$\delta\rho^{a} = D\varepsilon^{a} - 2\,(\varepsilon,\ R^{a}) \qquad (33)$$

$$\{\delta\rho = D\varepsilon - 2\,(\varepsilon,\ R)\}$$

where the curvature terms relate to ε^a, $\{\varepsilon\}$.

DYNAMICS

Gravitational theories follow from an Action of the form

$$A = \int_{\mathcal{M}^4} R^A \wedge \zeta_A \qquad (34)$$

where $\mathcal{M}^4$ is any 4-dimensional submanifold of G, and $\zeta_A = \zeta_A(\rho)$ is a 2-form constructed as a quadratic polynomial in the ρ^A. A should be stationary with respect to all variations of ρ^A and of $\mathcal{M}^4$. This last condition turns out, however, to be trivially satisfied by virtue of the general covariance of the theory. The theory should admit "flat" space $\rho^A = \omega^A$ as a solution. The field equations can be symbolically written as

$$R^A \wedge \frac{\delta\zeta_A}{\delta\rho^B} - D\zeta_B = 0 \qquad (35)$$

which is satisfied by $\rho^A = \omega^A$, $R^A = 0$ only if

$$D\zeta_A(\omega) = 0 \tag{36}$$

A form ζ_A satisfying (36) will be designated as a pseudo-curvature.

Clearly, the action (34) is not invariant under all G-gauge transformations. For example, Einstein gravity [10], which can be written as in (34),

$$\zeta_{ab} = \varepsilon_{abcd}\,\rho^c \wedge \rho^d , \qquad \zeta_a = 0$$
$$A = \frac{1}{8}\int_{\mathcal{M}^4} R^{ab} \wedge \zeta_{ab} \tag{37}$$

is only SO(3,1) gauge-invariant, (which was the reason for gauging AGCT or "parallel transport" rather than translations). Therefore, the assumption of a pseudo-curvature ζ_A breaks the G-symmetry.

The action (34) for supergravity is given by selecting ζ_A as:

$$\zeta_{ab} = \varepsilon_{abcd}\,\rho^c \wedge \rho^d$$
$$\zeta_a = 0 \tag{38}$$
$$\bar{\zeta}_\alpha = -4i\,\bar{\rho}\gamma^a\,\rho^a\gamma_5 \quad \text{or} \quad \zeta^\alpha = -4i\,\gamma_5\,\gamma^a\,\rho^a\,\rho$$

We assume the validity of the field-equations (35) throughout G. Clearly, if A is invariant under a gauge subgroup $H \subset G$ (i.e. SO(3,1) for both P and GP), a solution on G can be derived from its boundary value on $Z = G/_H$ through our factorization hypothesis (16), which we rewrite in simplified form

$$\rho^A = \omega_H^A + \tau^B\,\text{ad}\,(f^{-1})_B{}^{\cdot A} \tag{39}$$

where, locally, G has coordinates (z, f), $z\varepsilon Z$, $f\varepsilon H$, τ^B are the (factorized) forms on Z, ω_H^A the restrictions of the Cartan forms ω^A to H. We conjecture that in general ρ^A is determined by its boundary values on any $\mathcal{M}^4$, apart from a generic coordinate transformation on G. This can be proved to hold if a solution of (35) is sufficiently close to a factorized one. Globally, one cannot exclude the possible existence of "twisted" topologically inequivalent factorized solutions sharing the same boundary conditions on $\mathcal{M}^4$. Z is $\mathcal{M}^4$ in P, and $\mathcal{M}^{4/4}$ in GP. For P and Einstein's gravity, Z thus coincides with ^{4} so that (39) connects solutions on possible choices of $\mathcal{M}^4$ as discussed. For GP, $\mathcal{M}^4$ is still a submanifold of $\mathcal{M}^{4/4}$. In principle, eq. (35-36) should determine the extension of the form ρ^A to $\mathcal{M}^{4/4}$ from their restriction on $\mathcal{M}^4$. However, this is not achieved through an ordinary gauge transformation as it is for the Ξ^{ab} variables on SO(3,1). In order to compute the change in the ρ^A while moving infinitesimally from a surface $\mathcal{M}^4$ to $\mathcal{M}^{4'}$, we utilize the AGCT gauge of (32)

$$\delta\rho^A = D\,\varepsilon^A - 2\,(\varepsilon,\, R^A) \tag{32'}$$

(The ε^A are related to the lapse and shift functions of the standard canonical formalism of General Relativity.) For the index A in the Majorana range, these are the "local supersymmetry" transformations of supergravity. However, the exact comparison with supergravity requires the repeated use of field equations (35) - (36), yielding "Dynamically Restricted" AGCT. The field equations are,

$$R^a = 0 \tag{40a}$$

$$R^{ab}\ \rho^f\ \varepsilon_{abfe}\ \{-\ 2i\ \overline{R}\ \gamma_5\gamma_e\ \rho\} = 0 \tag{40b}$$

$$\{\gamma^a\rho^a\ R = 0\}$$

Direct calculation of the anholonomic components $R_{BC}{}^A$ from (40) yields,

$$R_{BC}{}^a = 0\ ,\qquad \{R_{\alpha\beta}{}^A = 0\}\ ,\qquad R_{(ab)\ B}{}^A = 0 \tag{41}$$

where a is a (translation) vector index, (ab) skew-symmetric tensor value (for an SO(3,1) direction) component, α, β are Majorana spinor indices for the odd generators. Inserting the surviving components in eq. (33) reproduces the conventional supergravity transformations[4)]

$$\delta\rho^{ab} = D\,\varepsilon^{ab} - \rho^c\,\varepsilon^d\,R_{cd}{}^{ab}\ \{-\ \rho^c\varepsilon^{\alpha}\ R_{c\alpha}{}^{ab}\} \tag{42a}$$

$$\delta\rho^a = D\,\varepsilon^a \tag{42b}$$

$$\{\delta\rho = D\,\varepsilon - \rho^c\ \varepsilon^d\ R_{cd}\} \tag{42c}$$

where D relates to P or GP and is given in (4). In this sense, the exponentiation of equations (33) starting from an arbitrary $\mathcal{M}^4$ is "condensed" into forms ρ^A on $\mathcal{M}^{4/4}$, and the $\mathcal{M}^{4/4}$ theory can be viewed as the collection of supersymmetric-related theories on $\mathcal{M}^4$. Formula (33) therefore solve in principle the problem of constructing supersymmetric transformations for any theory. Clearly, the field equations impose very stringent conditions on $R_{BE}{}^A$, and supergravity is certainly unique in yielding these restrictions. However, any covariant theory on GP or $\mathcal{M}^{4/4}$ is compatible with (33), with the specific form of $R_{BE}{}^A$ depending on the field equations. The values of these curvature components may then be much more complicated and there is no a priori guarantee that $R_{BE}{}^{A}$ are functionals of the restrictions of ρ^{A} on $\mathcal{M}^4$ only, as needed in order that (39) be an effective transformation. In the present formulation, the choice of an action is highly restricted by (36). Indeed, we have found in a recent study[11] with J. Thierry-Mieg that there exists no extended supergravity of type (34), with the ζ_{ab} of General Relativity, in which the internal symmetry is locally gauged. In all supersymmetry theories, we expect the $\mathcal{M}^4$ to remain an even 4-dimensional manifold. This avoids the use of the Berezin integration [8] on the variables, whose formal difficulties have impeded the development of viable 8-dimensional actions. From our point of view, the odd variables are just a shorthand for a collection of Fermi fields needed to specify $\mathcal{M}^4 \subset G$.

PSEUDO-INVARIANCE

A somewhat different point of view can be taken with respect to those transformations which require AGCT. J. Thierry-Mieg has recently [12] pointed out that since the equations of motion impose $R^a = 0$ (40a), we find in (42b) that the effective, dynamically restricted AGCT is finally identical to an ordinary translation gauge! In other words, the Lagrangian density is "pseudo-invariant" under a translation gauge, i.e. it becomes invariant after the application of the equations of motion. This was in fact the method used in deriving supergravity originally [4], except that the emergence of the (bracketed) supersymmetric transformation in (42a) is algebraically lost. Moreover, these pseudo-invariant translations cannot be exponentiated, since one has to apply the equations of motion after each infinitesimal step. No doubt that the AGCT gauge does provide a unique group-theoretical and geometrical justification for such transformations.

AFFINE THEORIES

There are several ways in which Gravity and Supergravity can be extended, besides inserting internal degrees of freedom as in "Extended Supergravity". The Poincaré group can be replaced by the de Sitter group; alternatively it can be extended to either the Conformal Group SU(2,2) or the Affine Group GA(4R). We have found it more promising to follow the latter path [13], leading to the Metric-Affine version of Gravity [14]. Both SU(2,2) and GA(4R) can be further extended into GLA. For the Affine Group, this involves an infinite GLA [15] resembling those of the Dual Models. With T. Sherry we have constructed such an algebra for GL(3R), and now hope to provide the relevant ones for GL(4R) and GA(4R), and investigate those theories.

Yet another modification involves replacing the Lagrangian in (34) by a quadratic expression in $R^A \wedge {}^*R_A$. Such Lagrangians have been suggested by Stephenson, Yang, Mansouri and Chang etc. However, since the real reason in using a quadratic Lagrangian is to recover the simple structure of Yang-Mills theories, we also have to see to the preservation of both gravitational gauge-fields, ρ^a and ρ^{ab}. General Relativity allows only one to propagate, since (34) contains only one Kinetic energy term, the $d\rho^{ab}_{ab}$ in R^{ab}. To preserve both fields we have to use R^a as well as R^{ab}, so that the Lagrangian will involve all 10 "curvatures" R^A (or 20 in GA(4R)). On the other hand, theories which allow such fields to propagate still have to meet the experimental tests of General Relativity. This can be achieved if the new components of the interaction turn out to be of a confining nature and can be assumed to be strong and short ranged (or, alternatively, if they be much weaker than Newtonian gravity). We have recently constructed such a model [16] in which "strong" gravity may occur naturally as a result of writing an "orthodox" gauge theory.

REFERENCES

1. C.N. Yang, Ann. N.Y. Acad. of Sc. 292, 86 (1977), expecially pp. 92-97.
2. F.W. Hehl, P. Von der Heyde, G.D. Kerlick, J.M. Nester, Rev. Mod. Phys. 48, 393 (1976).
3. Y. Ne'eman, in Differential Geom. Meth, in Phys. (Bonn 1977), K. Bleuler and A. Reetz editors, Springer Verlag (in print).
4. D.Z. Freedman, P. van Nieuwenhuizen and S. Ferrara, Phys. Rev. D13, 3214 (1976).
5. J. Wess and B. Zumino, Phys. Letters 66B, 361 (1977).
6. Y. Ne'eman and T. Regge, Physics Letters (1978); also Rivista del Nuovo Cimento, to be published.
7. Yu. A. Golfand and E.P. Likhtman, JETP Letters 13, 452 (1971); J. Wess and B. Zumino, Nucl. Phys. B70, 39 (1974).
8. F.A. Berezin and G.I. Katz, Mat. Sb. (USSR) 82, 124 (1970), (English translation, 11, 311, 1970).
9. L. Corwin, Y. Ne'eman and S. Sternberg, Rev. Mod. Phys. 47, 573 (1975).
10. A. Trautman, Bull. Acad. Pol. Sci., Sen. Sci. Matt. Astr. Phys.
11. Y. Ne'eman and J. Thierry-Mieg, to be pub.
12. J. Thierry-Mieg, to be pub.
13. F.W. Hehl, E.A. Lord and Y. Ne'eman, Phys. Lett. 71B, 432 (1977); also Phys. Rev. D17, 428 (1978).
14. F.W. Hehl, G.D. Kerlich and P.V. d. Heyde, Phys. Lett. 63B, 446 (1976).
15. Y. Ne'eman and T. Sherry, Phys. Lett. (1978).
16. F.W. Hchl, Y. Ne'eman, J. Nitsch and P.V. d. Heyde, to be pub.

ENVOI

It seems highly fitting that Dirac's and Dirac's equation's anniversary should include a discussion of Supersymmetry, Supergravity and Affine (World) Band-Spinors. These three represent evolutionary progeny of the Dirac Spinor, extending respectively the Poincaré Algebra Generators, Gravity and World-Tensors.

SPIN, SUPERSYMMETRY, AND SQUARE ROOTS OF CONSTRAINTS*

Claudio Teitelboim†
The Institute for Advanced Study, Princeton, New Jersey 08540
and
Joseph Henry Laboratories
Princeton University, Princeton, New Jersey 08540

ABSTRACT

It is argued that the natural way of introducing spin degrees of a freedom in a generally covariant system is to take the square root, à la Dirac, of the first class Hamiltonian constraints which generate the dynamical evolution of the system without spin. The theories obtained in this way are locally supersymmetric by construction. A comparative analysis of the relativistic spinning particle and of supergravity is given, which shows that supergravity bears to ordinary general relativity the same relation that a Dirac particle bears to a Klein Gordon particle.

. . . *"I noticed that if you form the expression* $\sigma_1 p_1 + \sigma_2 p_2 + \sigma_3 p_3$ *and squared it,* p_1, p_2, p_3 *being the three components of momentum, you got just* $p_1^2 + p_2^2 + p_3^2$ *, the square of the momentum. This was a very pretty mathematical result. I was quite excited over it. It seemed that it must be of importance.* . . . *"*

P.A.M. DIRAC, in "Recollections of an Exciting Era"[1]

It is a great pleasure and privilege to report at this conference in honor of Professor Dirac and of the fifty years of the Dirac equation. It is all the more so, as the work which I would like to summarize[2,3] is a direct offspring of Dirac's way of obtaining the relativistic wave equation for the electron as a square root of the Klein-Gordon equation.

The manner in which Dirac added spin degrees of freedom to a point particle in 1928 is indeed a remarkable one. He replaced the quantum mechanical equation for a particle with no spin by one richer in structure but, at the same time, more elementary. The new equation was richer in structure because it contained additional dynami-

*Work supported in part by National Science Foundation Grants No.PHY-77-20612 to the Institute for Advanced Study and PHY76-82662 to Princeton University.

†Alfred P. Sloan Research Fellow

ISSN: 0094-243X/78/134/$1.50

cal variables (the spin) and was more elementary because used <u>twice</u> (i.e. squared) gave back the original equation.

It now appears that one can use Dirac's square root procedure to introduce spin degrees of freedom in a natural way not only on a system devoid of spatial extension (a point particle), but also on extended systems such as a string[4] (dimension one), a membrane[5] (dimension two) and a three dimensional space. The resulting theories turn out to be in each case "locally supersymmetric" in today's parlance[6] and the three dimensional case yields supergravity theory.[7] Thus in retrospect the Dirac electron emerges in a precise sense as a supersymmetric system.[8] Therefore, besides the fifty years of the electron wave equation, we also celebrate today fifty years of supersymmetry!

The extension of the square root idea to more complex systems makes essential use of another important development of Dirac's, namely his generalization of the mechanics of Hamilton to encompass systems with constraints.[9] In effect, the object of which one takes the square root of is an operator which when applied on physically allowed states yields zero. In Dirac's terminology such an operator is called a first class constraint. (Note, incidentally, that this means that one is taking the square root of the Schroedinger equations of motion and not of the Heisenberg ones).

Now, it is a general feature of constrained Hamiltonian systems, that for each independent first class constraint there appears one arbitrary function of time in the solution of the equations of motion. Or, in other words, each first class constraint generates a "gauge transformation". In the particle case the constraint $p_\mu p^\mu + m^2 = 0$ generates displacements along its world line, and the corresponding arbitrary function may be taken to be the rate of change of the coordinate time z^0 with respect to an arbitrary parameter τ used to label the worldline. The "gauge transformation" is in this case a worldline reparametrization, and the invariance of the equations of motion (or of the action) under the transformation may be appropriately termed "general covariance".

It suggests itself at this point to extend Dirac's idea and to attempt to introduce spin on systems with more structure than a point particle by taking the square root of the first class constraints which generate the dynamical evolution of the system without spin. In doing so one should restrict oneself to generally covariant systems, as only for those is the dynamical evolution generated by a first class constraint. Moreover it turns out that as a rule, such systems (or at least those among them for which the action has a direct geometrical significance) possess at least one constraint quadratic in the momenta, which makes the square root approach not obviously impracticable.

After the square roots are taken the total number of constraints of the theory increases (but also new dynamical variables are introduced). Besides the constraints of the original theory there appear now new constraints (the square roots) which are closed under anticommutation ("Fermi functions"). The complete set of constraints closes then according to a graded algebra. If interpreted as generators of a symmetry transformation the Fermi constraints mix Bose variables (obeying commutation relations) with the newly introduced Fermi variables (obeying anticommutation relations) which shows that the theory is supersymmetric.[6] Hence from this point of view supersymmetry and the introduction of spin are both consequences of the square root process.

As a last general point I would like to remark that in every known case but the free particle (which is abnormally simple) the Bose constraints of the new theory differ in general from those which one started with before taking the square root. The difference is due to the presence of terms containing the new spin degrees of freedom. In the classical limit the Fermi degrees of freedom disappear and one requires, as part of the definition of an acceptable square root, that in that limit the Bose constraints should reduce to the previous ones and the Fermi constraints should vanish identically. Therefore the theory reduces to the original one when the new purely quantum mechanical features introduced by the square root process are neglected. The demand that the new theory must reduce to the original one in the classical limit poses a strong restriction on the allowed square roots and, together with the requirement that the constraint algebra be closed and some natural simplicity assumptions, fixes uniquely in practice the new system of constraints.

I should stress here that although they are classically absent the extra terms are quite important physically. For example in the case of a particle in an electromagnetic field the extra term is $\sigma_{\mu\nu}F^{\mu\nu}$ which yields the Zeeman effect. In the case of supergravity the Fermionic contribution is the energy-momentum density of the spin three halves field.

To make these remarks more tangible I would like to finish by sketching how supergravity theory can be regarded as the square root of the standard Hamiltonian form of general relativity (also worked out by Dirac[10], and independently by Arnowitt, Deser, and Misner[11]). In order to bring out the analogy with the particle case as clearly as possible I will present the discussion in the form of comparative tables. Many technical points will be omitted. A more detailed treatment may be found in Refs. 2 and 3.

Table I Original theory (without spin)

Dynamical system	Spinless point	"Spinless three-dimensional space"
Theory	Free spinless relativistic particle (Klein-Gordon)	Einstein's general relativity (without matter)
Canonical variables	$z^{\mu}(\tau)$, $p_{\mu}(\tau)$ (position and conjugate momentum)	$g_{ij}(\underset{\sim}{x},\tau)$, $\pi^{ij}(\underset{\sim}{x},\tau)$ (three-geometry and conjugate momentum)

Constraints	$\mathcal{H} = \eta^{\mu\nu}p_\mu p_\nu + m^2$ Generator of displacements along world line. One in number $\eta^{\mu\nu} = (-1,+1,+1,+1)$ (Minkowski metric)	$\mathcal{H} = G_{ijkl}\pi^{ij}\pi^{kl} - g^{\frac{1}{2}}R$ $\mathcal{H}_i = -2\nabla_j\pi^j{}_i$ Generators of hypersurface deformations in spacetime. Four in number for each space point $G_{ijkl} = \frac{1}{2}g^{-\frac{1}{2}}(g_{ik}g_{jl} + g_{il}g_{jk} - g_{ij}g_{kl})$ (Supermetric)
Hamiltonian	$H = N(\tau)\mathcal{H}$ $N(\tau)$ arbitrary function	$H = \int d^3x(N^\perp\mathcal{H}_\perp + N^i\mathcal{H}_i)$ $N^\perp(\underset{\sim}{x},\tau), N^i(\underset{\sim}{x},\tau)$ arbitrary functions
Schroedinger equations of motion	$\dot{F} = [F,H]$ for any dynamical variable F. Second order ordinary differential equations for z^μ (free particle equations) if the momenta are eliminated	$\dot{F} = [F,H]$ for any dynamical variable F. Second order partial differential equations for $g_{\mu\nu}$ (Einstein's equations) if the momenta are eliminated.
Quantum Equations of motion in Schroedinger picture	$\mathcal{H}\|\Psi\rangle = 0$ where $p_\mu = \frac{1}{i}\frac{\partial}{\partial z^\mu}$ Second order partial differential equation in z^μ	$\mathcal{H}_\mu(\underset{\sim}{x})\|\Psi\rangle = 0$ where $\pi^{ij} = \frac{1}{i}\frac{\delta}{\delta g_{ij}}$ System of second order functional differential equations in g_{ij}.

Table II Square root theory (with spin)

Dynamical system	"Spinning point"	"Spinning three-dimensional space"
Theory	Free spinning relativistic particle (Dirac)	Supergravity
Canonical variables	$z^\mu(\tau), p_\mu(\tau)$, $\theta_\mu(\tau), \theta_5(\tau)$	$g_{ij}(\underset{\sim}{x},\tau), \pi^{ij}(\underset{\sim}{x},\tau)$ $\psi_i{}^A(\underset{\sim}{x},\tau)$

Equal time anticommutation rules of new (Fermi) variables fixed by "square root requirement"	$\{\theta_\mu,\theta_\nu\} = \eta_{\mu\nu}$ $\{\theta_5,\theta_5\} = 1$ $\{\theta_\mu,\theta_5\} = 0$	$\left\{\psi_i^A(\underset{\sim}{x}),\psi_j^B(\underset{\sim}{x}')\right\} =$ $= \left(\alpha_i\alpha_j\right)^{BA}\delta(\underset{\sim}{x},\underset{\sim}{x}')$
Usual name of Fermi variables	Dirac matrices	Vector spinor
New Fermi constant (Supersymmetry generator)	$\mathcal{S} = \theta_\mu p^\mu + m\theta_5$	$\mathcal{S} = 2\alpha_i\psi_j\pi^{ij}$+"curl$\psi$"
Infinitesimal supersymmetry transformation on sample dynamical variable	$\delta z^\mu = [z^\mu,\varepsilon\mathcal{S}]$ $= i\varepsilon\theta^\mu$	$\delta g_{ij} = [g_{ij},\int d^3x\varepsilon_A(\underset{\sim}{x})\mathcal{S}^A(\underset{\sim}{x})]$ $= i\varepsilon(\alpha_i\psi_j + \alpha_j\psi_i)$
Square root property	$\{\mathcal{S},\mathcal{S}\} = \mathcal{H}$	$\{\mathcal{S}^A(\underset{\sim}{x}),\mathcal{S}^B(\underset{\sim}{x}')\} =$ $=\left(\delta^{AB}\mathcal{H}_\perp + \alpha^i\mathcal{H}_i\right)\delta(\underset{\sim}{x},\underset{\sim}{x}')$
Constraints	$\mathcal{S}$, $\mathcal{H}$ (two in number)	$\mathcal{S}^A(\underset{\sim}{x})$, $\mathcal{H}_\perp(\underset{\sim}{x})$, $\mathcal{H}_i(\underset{\sim}{x})$ (four in number per space point)
Hamiltonian	$H = N(\tau)\mathcal{H}(\tau)+\varepsilon(\tau)\mathcal{S}(\tau)$ N, ε arbitrary functions	$H = \int d^3x\left(N^\perp\mathcal{H}_\perp + N^i\mathcal{H}_i + \varepsilon^A\mathcal{S}_A\right)$ $N^\perp$, $(\underset{\sim}{x},t)$, $N^i(\underset{\sim}{x},t)$, $\varepsilon(\underset{\sim}{x},t)$ arbitrary functions
Heisenberg equations of motion	$\dot{F} = [F,H]$ for any dynamical variable F. Second order ordinary differential equations for z^μ (if momenta are eliminated) and first order ordinary differential equations for θ_μ,θ_5.	$\dot{F} = [F,H]$ for any dynamical variable F. Second order partial differential equations for $g_{\mu\nu}$ (if momenta are eliminated) and first order partial differential equations for ψ_i

Schroedinger equations of motion	$\mathcal{S}\vert\Psi\rangle = 0$ First order partial differential equation in z^{μ}. It implies $\mathcal{H}\vert\Psi\rangle = 0$ by the square root property.	$\mathcal{S}^{A}(\underset{\sim}{x})\vert\Psi\rangle = 0$ System of first order functional differential equations in g_{ij}. They imply $\mathcal{H}_{\perp}\vert\psi\rangle = 0$ and $\mathcal{H}_{i}\vert\psi\rangle = 0$ by the square root property.

It has been said here that after fifty years all is well with the Dirac equation. I hope that the preceding discussion conveys, at least in some measure, that in addition it continues to lead us through new and fascinating avenues of research.

REFERENCES

1. P.A.M. Dirac in History of Twentieth Century Physics, Proceedings of the Varenna Summer School, LVII Corso (Academic, N.Y., 1977), p. 142.
2. C. Teitelboim, Phys. Rev. Lett. 38, 1106 (1977).
3. R. Tabensky and C. Teitelboim, Phys. Lett. 69B, 453 (1977).
4. Y. Nambu, lectures in Copenhagen, 1970; T. Goto, Prog. Theor. Phys. 46, 1560 (1971); P. Ramond, Phys. Rev. D3, 2415 (1971); A. Neveu and J. H. Schwarz, Nucl. Phys. B31, 86 (1971); S. Deser and B. Zumino, Phys. Lett. 65B, 369 (1976); L. Brink, P. Di Vecchia and P. Howe, Phys. Lett. 65B, 471 (1976).
5. P. A. Collins and R. W. Tucker, Nucl. Phys. B112, 150 (1976), P.S. Howe and R.W. Tucker, J. Math. Phys. 19, 869 (1978), 19, 981 (1978). The Hamiltonian for the "spinning membrane" has been worked out by M. Pilati (unpublished).
6. D. V. Volkov and V. A. Soroka, JEPT Letters 18, 529 (1973); J. Wess and B. Zumino, Nucl. Phys. B70, 39 (1974); Phys. Lett. 49B, 52 (1974).
7. D. Z. Freedman, P. van Nieuwenhuizen and S. Ferrara, Phys. Rev. D13, 3214 (1976); S. Deser and B. Zumino, Phys. Lett. 62B, 335 (1976); D. Z. Freedman and P. van Nieuwenhuizen, Phys. Rev. D14, 912 (1976).
8. A. Barducci, R. Casalbuoni and L. Lusanna, Nuovo Cimento 35A, 377 (1976); F. A. Berezin and M. S. Marinov, Ann. Phys. (N.Y.) 104, 336 (1977); L. Brink, S. Deser, B. Zumino and P. Howe, Phys. Lett. 64B, 435 (1976); L. Brink, P. di Vecchia and P. Howe, Nucl. Phys. B118, 76 (1977); C. Bachas, Phys. Lett., in press.
9. P.A.M. Dirac, Can. J. Math. 2, 129 (1950); 3, 1 (1951); Lectures on Quantum Mechanics (Academic, N. Y., 1964).
10. P.A.M. Dirac, Proc. Roy. Soc. (London) A246, 326 (1958).
11. R. Arnowitt, S. Deser and C. W. Misner, in Gravitation, an Introduction to Current Research, ed. L. Witten (Wiley, N.Y., 1962)

GRAVITATIONAL THEORIES GENERATED BY GROUPS OF TRANSFORMATION AND DIRAC'S LARGE NUMBER HYPOTHESIS.

by

Leopold Halpern
Florida State University
Tallahassee, Florida 32306

ABSTRACT

Two different methods are presented by which gravitational theories can be developed from intransitive groups of transformations. A minimal invariant variety of the group forms the unperturbed universe. A Lie derivative of spinors is defined in this space-time. A condition is derived for the coincidence of individual group trajectories and geodesics. A speculation on the physical meaning also of nongeodesic group trajectories is put forward. The gravitational fields of localized masses break the homogenous character of the space. This can be achieved either by introducing a metric with suitable homogenuity conditions into the higher dimensional space which is transformed by the group, or alternatively by applying a gauge formalism to the field of basis vectors of the group. The relation of the second method with the metric is shown; It results in field equations of the Maxwell-Yang Mills type corresponding to and generalizing the metric Lagrangian: $\sqrt{g}\, R_{\mu\nu\rho\sigma} R^{\mu\nu\rho\sigma}$

It is shown that a theory which fulfills the requirement of Dirac's Large Number Hypothesis can not be based on an invariance group with Killing vectors with time components. Some aspects of an approach to such a theory are discussed.

Detailed version of the lecture presented at the meeting in honour of Dirac's 75th birthday in Tallahassee April 6-8, 1978.

ISSN: 0094-243X/78/140/$1.50

I. INTRODUCTION

Einstein's general theory of relativity derives the local gravitational law and the space time structure of the whole universe from field equations with boundary and initial conditions.

Dirac's large number hypothesis assumes a relation between all the dimensionless large number parameters in physics. Expressed in units of magnitude characteristic for elementary particle physics, Newton's gravitational constant $G \approx 10^{-39}$. The present age (and radius) of the expanding universe $t \approx G^{-1}$ and the number of massive fermions in this universe $N \approx G^{-2}$. Dirac conjectures that as t increases, G and N vary accordingly. Einstein's theory is not compatible with these assumptions. Einstein himself[1] and later Jordan[2] and Brans and Dicke[3] considered modifications of general relativity in which G is a scalar field which varies in accord with the universe, but the results were not confirmed by observation.

Dirac[4] introduced a theory in which general relativity is rigorously maintained in a suitable gauge (called mechanical gauge) and his large number hypothesis is true in a different gauge (atomic gauge) in which the quantum of action appears. A conformal transformation relates the metrics of the two gauges in the simplest case.[5] Dirac recently modified his views. He gave up matter creation in favor of matter accumulation[6] similar to his original views.[7] The state of possible verifications of the large number hypothesis (LNH) has been the subject of a workshop at Florida State University.[8]

I shall present here a different approach to modifications of general relativity which should lead to the LNH.

I have shown previously that metric theories closely related to general relativity can be obtained by the requirement that in the local limit the metric be transformable into that of an invariant variety of a group of transformation other than the

Poincaré group,[9*] in particular the De Sitter group. The invariant varieties of the invariance group may determine the global topology of the universe and the gravitational law determines the local metric. One may consider here a local law which acts only for concentrations of matter of density different from the average in the universe or a global law where the topology of the universe is determined by the average matter distribution. The latter case leads then to a complementarity between the chosen gravitational law and the local invariance group somewhat analog to the complementarity of physics and geometry discovered by Riemann.

We shall see in the following that Dirac's LNH which does not allow constant dimensionless large numbers, excludes universes which are invariant varieties of groups of transformations with Killing vectors that have time components in a preferred frame, but that without such invariance groups a suitable gravitational law may be formulated.

II.
A GENERALIZATION OF DIRAC'S METHOD TO OBTAIN GROUP COVARIANT FIELD EQUATIONS

We consider a continuous group of transformations G_r of a n-dimensional space with r essential parameters.** The rank of the matrix of symbols of the group: (ζ^i_α) with ($i=1\ldots n$, $\alpha=1\ldots r$) be $q<n<r$, so that the group is intransitive. There exist then q-dimensional invariant varieties.

One can in general find a metric of the n-dimensional space, such that G_r is a group of motions in it and each of the q-dimensional minimal invariant varieties is a Riemannian subspace imbedded in the n-dimensional space. We are interested in the case $q=4$ and signature+2, when such an invariant variety may be considered as the space-time of a universe with group of motion G_r. One can then introduce a coordinate system such that everywhere in the V_n:

$$\xi^h_\alpha = 0 , \quad g_{hi} = 0 \ (i \neq h) , \quad g_{hh} = \pm 1 \tag{1}$$
$$(\alpha = 1 \ldots r , \ h = q+1 \ldots n , \ i = 1 \ldots n)$$

as follows from the Killing equations:

$$\frac{\partial g_{jk}}{\partial x^l} \xi^l_\alpha + g_{jl} \frac{\partial \xi^l_\alpha}{\partial x^k} + g_{lk} \frac{\partial \xi^l_\alpha}{\partial x^j} = 0 \tag{2}$$

Partial differential equations involving tensor fields in V_n, which are of covariant form w.r.t. general coordinate transformations can be expressed on any invariant variety V_q if the following conditions are fulfilled: 1. All tensor components with index $i>q$ vanish in our coordinate system.[13] 2. All tensor components and also the metric tensor are homogenous functions of the variables $x^{q+1} \ldots x^n$.

*Simultaneously with the mentioned work,[9] McDowell and Mansouri suggested a supersymmetric De Sitter invariant gravitational theory.[10]
**Our notation follows as far as possible that of L.P. Eisenhart.[11,12]

Another way to obtain covariant expressions w.r.t. G_r from such tensor fields on V_q if condition 1 is fulfilled, is to form instead of the covariant-metric derivatives of the tensor field the Lie derivatives w.r.t. the symbols ζ^i_α. With a nonsingular metric $\gamma_{\alpha\beta}$ in group space one can also form covariant differential equations on V_q from such operations.

I define in the next section also the Lie derivative of a spinor in the space where G_r is a group of motion.

The method outlined in the present section is a generalization of Dirac's method to obtain DeSitter and conformally covariant field equations.[13,14] Some physical aspects of the results are presented in Section IV and the generalization to the presence of gravitational fields is performed in Section V.

III. THE LIE DERIVATIVE OF A SPINOR

The metric derivative of a spinor covariant w.r.t. general coordinate and spin transformation is:

$$\psi_{;k} = \frac{\partial\psi}{\partial x^k} + \Gamma_k\psi \tag{1}$$

where the metric spinor connection Γ_k is defined by the relation:

$$\frac{\partial\gamma^i}{\partial x^k} + \gamma^\ell\left\{ {i \atop \ell\, k} \right\} - [\gamma^i, \Gamma_k] = 0 \tag{1a}$$

in the Vierbein notation:

$$\gamma^k = h^k_{a|}\,\overset{\circ}{\gamma}{}^a, \qquad h^k_{a|}\,h^\ell_{b|}\,\overset{\circ}{g}{}^{ab} = g^{k\ell}$$
$$\{\overset{\circ}{\gamma}{}^a, \overset{\circ}{\gamma}{}^b\} = 2\overset{\circ}{g}{}^{ab}, \qquad \{\gamma^k, \gamma^\ell\} = 2g^{k\ell} \tag{2}$$

$$\Gamma_k = \frac{1}{4} h_{c|}{}^m\, h_{d|m;k}\,\overset{\circ}{G}{}^{cd} \qquad \left(\overset{\circ}{G}{}^{cd} = \tfrac{1}{2}[\overset{\circ}{\gamma}{}^c, \overset{\circ}{\gamma}{}^d]\right) \tag{2a}$$

On a space where G_r is a group of motion, I define the Lie derivative of a spinor w.r.t. ζ^i_α:

$$\psi_{\|\alpha} = \frac{\partial\psi}{\partial x^\ell}\zeta^\ell_\alpha + \Gamma^L_\alpha\psi \tag{3}$$

with the connection Γ^L_α defined by:

$$\frac{\partial\gamma^i}{\partial x^\ell}\zeta^\ell_\alpha - \frac{\partial\zeta^i_\alpha}{\partial x^\ell}\gamma^\ell - [\gamma^i, \Gamma^L_\alpha] = 0 \tag{3a}$$

One can supplement ζ^i_α by q-1 linear independent symbols of the group and form their algebraic complements ζ^β_k:

$$\zeta^i_\alpha\zeta^\alpha_k = \delta^i_k, \qquad \zeta^k_\alpha\zeta^\beta_k = \delta^\beta_\alpha \qquad (i,k,\alpha,\beta = 1\dots q) \tag{4}$$

for the remaining symbols we have:

$$\zeta^i_\beta = \varphi^\alpha_\beta(x)\,\zeta^i_\alpha \qquad \left({\alpha = 1\dots q \atop \beta = q+1\dots r}\right) \tag{5}$$

One can now form the nonsymmetric connection:[15]

$$\Lambda^{i}_{k\ell} = -\xi^{\alpha}_{\ell}\frac{\partial \xi^{i}_{\alpha}}{\partial x^{k}} = \xi^{i}_{\alpha}\frac{\partial \xi^{\alpha}_{\ell}}{\partial x^{k}} \qquad (\alpha, \ell = 1 \ldots q) \text{ sum over } \alpha \tag{6}$$

and finds the integrability condition for Eq. (3a):

$$\frac{\partial \Gamma^{L}_{k}}{\partial x^{\ell}} - \frac{\partial \Gamma^{L}_{\ell}}{\partial x^{k}} + [\Gamma^{L}_{k}, \Gamma^{L}_{\ell}] = G^{rs}\Lambda_{rs\ell k} + \frac{\partial a_{k}}{\partial x^{\ell}} - \frac{\partial a_{\ell}}{\partial x^{k}} \tag{7}$$

here $\Gamma^{L}_{k} = \Gamma^{L}_{\alpha}\zeta^{\alpha}_{k}$ and

$$\Lambda_{rs\ell k} = g_{ri}\left(\frac{\partial \Lambda^{i}_{sk}}{\partial x^{\ell}} - \frac{\partial \Lambda^{i}_{s\ell}}{\partial x^{k}} + \Lambda^{j}_{sk}\Lambda^{i}_{j\ell} - \Lambda^{j}_{s\ell}\Lambda^{i}_{jk}\right) \tag{7a}$$

The integrability condition can only be fulfilled if

$$\Lambda_{rs\ell k} + \Lambda_{sr\ell k} = 0 \tag{7b}$$

but this relation can be shown to be a consequence of G_r being a group of motion.[16]

Using Γ^{L}_{k} as connection in the spinning electron equation covariant w.r.t. G_r, one can also show that the current is conserved. In this spinning electron equation γ^{k} are replaced by γ^{β}, the generators of G_r expressed by the γ^{n} (if such generators can be formed) see Ref. 13,9.

The two connections are related as follows:

$$\Gamma^{L}_{\alpha} - \Gamma_{k}\xi^{k}_{\alpha} = \frac{1}{8}G^{ik}\left(\frac{\partial}{\partial x^{i}}\xi_{\alpha k} - \frac{\partial}{\partial x^{k}}\xi_{\alpha i}\right) \tag{8}$$

In case of generalized Lorentz and rotation groups, the last term has itself the form of a generator in spin space and the contribution of such a term to the Dirac equation is thus only a mass term of order of magnitude determined by the inverse of the radius of the homogenous space, which one may identify with the universe.

IV. A THEOREM ON GROUP TRAJECTORIES AND GEODESICS. SPECIALIZATION ON THE GROUP COVARIANT LAW OF MOTION

The following theorem is a generalization of theorems which apply to the geometry of the group space.[11] It applies here to the invariant variety of our intransitive group.

Theorem:

The trajectory of the symbol ξ^i_α of a group G_r on its minimal invariant variety V_q coincides with a geodesic of the Riemannian metric on V_q for which G_r is a group of motion, iff on every one of its points, there exist q linear independent symbols ξ^i_δ (e.g. $\delta = 1 \ldots 4$) (one of which may be ξ^i_α itself) such that:

$$[\xi_\alpha, \xi_\delta]^i g_{ik} \xi^k_\alpha = c^\gamma_{\alpha\delta} \xi^i_\gamma g_{ik} \xi^k_\alpha = 0 \qquad (\gamma = 1 \ldots r,\ \delta = 1 \ldots q) \qquad (1)$$

Proof: Every symbol fulfills Killing's equation (II.2).

$$\frac{\partial}{\partial x^\ell} g_{ik} \xi^\ell_\delta + g_{\ell k} \frac{\partial \xi^\ell_\delta}{\partial x^i} + g_{i\ell} \frac{\partial \xi^\ell_\delta}{\partial x^k} = 0$$

contracting the indices i and k with ξ_α and making use of Eq. (1).

$$g_{i\ell} \frac{\partial \xi^\ell_\delta}{\partial x^k} \xi^k_\alpha \xi^i_\alpha = g_{i\ell} \frac{\partial \xi^\ell_\alpha}{\partial x^k} \xi^k_\delta \xi^i_\alpha$$

results in:

$$\frac{\partial}{\partial x^\ell} (g_{ik} \xi^i_\alpha \xi^k_\alpha) \xi^\ell_\delta = 0 \qquad (2)$$

so that all derivatives of $g_{ik} \xi^i_\alpha \xi^k_\alpha$ vanish on the trajectory. One can choose the parameter for the trajectory such that $\dot{x}^k = \xi^k_\alpha(x)$ and thus $\ddot{x}^k = \frac{\partial}{\partial x^\ell}(\xi^k_\alpha) \xi^\ell_\alpha$

futhermore

$$\xi^\ell_\alpha \xi^m_\alpha \left\{ {k \atop \ell m} \right\} = \frac{1}{2} g^{ki} \left(2 \frac{\partial}{\partial x^m} g_{\ell i} - \frac{\partial}{\partial x^i} g_{\ell m}\right) \xi^\ell_\alpha \xi^m_\alpha$$

the right hand side because of Eq. (2) equals

$$g^{ki} \left(\frac{\partial}{\partial x^m} g_{\ell i} \xi^\ell_\alpha \xi^m_\alpha + g_{\ell m} \xi^\ell_\alpha \frac{\partial}{\partial x^i} \xi^m_\alpha\right) \qquad \text{which}$$

because of Killing's equation (II.2)

equals $-\frac{\partial}{\partial x^\ell}(\xi^k_\alpha) \xi^\ell_\alpha$ such that the equation of the geodesic for our parameter;

$$\ddot{x}^k + \left\{ {k \atop \ell m} \right\} \dot{x}^\ell \dot{x}^m = 0$$

is fulfilled.

What is the law of motion of a macroscopic body on this manifold? Gursey[17] in his review article on the De Sitter group points out that all the geodesics in this case are also trajectories of the group and he conjectures the motion along a time like geodesic.

One should request also for the general case that all the time like geodesics be identical or at least very well approximated by group trajectories. But even in case of the De Sitter group there are other trajectories (which do not correspond to maximal circles). Must we exclude such trajectories from the law of motion? I don't think so if we consider invariant varieties as large as the universe! A motion which does not approximate well the analog of a maximal circle as trajectory in De Sitter space for example, may be extremely rare for statistical reasons - Just as we rarely can find a macroscopic system violating the second fundamental theorem of thermodynamics. The phase space of a trajectory approximating a circle of radius comparable to our solar system would be so much smaller than that approximating the radius of the Universe!

Does not the wave equation on the invariant variety constructed according to the rules outlines in Section II take account of all the degrees of freedom of the group? A detailed discussion of this conjecture will be presented in a separate publication.

V. INTRODUCING A GENERAL METRIC

The space V_n on which the group G_r acts does not have a metric, but it is in general possible to introduce a metric for which G_r is a group of motion.[11]

A metric which fulfills:

$$\frac{\partial}{\partial x^l} g_{ik} \xi^l_\alpha + g_{lk} \frac{\partial}{\partial x^i} \xi^l_\alpha + g_{il} \frac{\partial}{\partial x^k} \xi^l_\alpha = 0 \tag{1a}$$

$$g_{iu} = 0 \quad g_{uv} = \mathring{g}_{uv} \quad (i,k,l = 1 \ldots q) \quad (\alpha = 1 \ldots r) \tag{1b}$$

$$(u,v = q+1 \ldots n) \quad \mathring{g}_{uv} = \text{flat sp. metric}$$

where x^i are coordinates of the minimal invariant varieties V_q, satisfies this condition. Choosing the g_{ik} of the V_q homogenous functions of second degree in the perpendicular coordinates X^u, makes all perpendicular components of the Riemann tensor vanish. This allows the interpretation of V_q as unperturbed space-time if $q=4$.

In case of a localized mass distribution, not all conditions (1a) can be fulfilled, but the remaining conditions can remain valid. One may postulate that the n-dimensional Einstein tensor vanishes in the free space surrounding these masses. This leads to a gravitational law that differs from Einsteins only by a cosmological member. I have shown this in detail in case of De Sitter symmetry.[9]

VI. INTRODUCING GRAVITATION WITHOUT METRIC AS GAUGE THEORY OF THE GROUP.

There have been numerous suggestions to introduce gravitation as a gauge field since Utiyana[18] applied the formalism of Yang Mills to the Lorentz group.

The present method operates directly on the mathematical framework and can thus be applied to a variety of invariance groups. It makes use of the covariance of the formalism of Lie groups w.r.t. linear transformations of the basis vectors. The generality of the formalism may be reduced by limiting the linear transformation to those of the adjoint group. We use however, the full group of linear transformations.

The intransitivity of the group G_r allows to perform a linear transformation S of the basis vectors ξ^i_β such that at any given ordinary point O, r-q of them, let us say $\beta = q+1..r$ are zero and the first q form there a complete system. The theory of Lie groups is formulated covariantly w.r.t. such a transformations.

I consider now an independent transformation of this kind at every point of V_q. The covariance of the formalism can then be restored by introducing a gauge potential $A_k(x)$ which is an r-dimensional square matrix in the group indices.

The derivative of the basis vectors is replaced by the invariant derivative:

$$\xi^i_{\alpha\cdot k} = \frac{\partial \xi^i_\alpha}{\partial x^k} + A^\beta_{\alpha k}(x)\, \xi^i_\beta \tag{1}$$

A linear transformation S of the basis vectors transforms

$$A_k \rightarrow S A_k S^{-1} - \frac{\partial S}{\partial x^k} S^{-1} \tag{2}$$

so that $\xi^i_{\alpha\cdot k}$ transforms by S as ξ^i_α.

One can now perform a linear transformation at every point of our minimal invariant variety V_q such that:

$$\xi^i_\alpha = \delta^i_\alpha \quad (\alpha = 1 \cdots q), \quad \xi^i_\beta = 0 \quad (\beta = q+1 \cdots r) \tag{3}$$

this transformation creates a gauge potential A_k which allows to express the Lie derivatives in a covariant way e.g., for a vector B:

$$B^i_{,\alpha} = \frac{\partial B^i}{\partial x^k} \xi^k_\alpha + \xi^i_\beta A^\beta_{\alpha k} B^k$$

$$\xi^i_\beta = \delta^i_\beta \qquad (\beta = 1 \cdots q) \tag{1a}$$

The gauge field F_{ik} transforms:

$$F_{ik} = \frac{\partial}{\partial x^i} A_k - \frac{\partial}{\partial x^k} A_i + [A_i, A_k] , \qquad F_{ik} \rightarrow S F_{ik} S^{-1} \tag{2a}$$

and can thus not be created or destroyed by the transformations. The field on our invariant variety V_q, with the metric for which G_r is a group of motion, vanishes.

The breaking of this symmetry results in gravitational fields characterized by nonvanishing F_{ik} and simultaneously by a different metric on V_q.

The relation between gauge field and metric I generalize as follows:

The metric fulfills the Killing equations expressed in covariant form w.r.t. gauge transformations with the generalized gauge potential.*

In the special gauge defined in Eq (3) the Killing equations (II.2) are modified to:

$$\frac{\partial g_{rs}}{\partial x^k} \delta^k_\alpha + g_{\ell s} \delta^\ell_\beta A^\beta_{\alpha r} + g_{r\ell} \delta^\ell_\beta A^\beta_{\alpha s} = 0 \tag{4}$$

(Greek indices 1...r, Latin indices 1...q)

The gauge potential by the relation (4) can thus characterize the metric and vice versa. Limited space prevents us to expand the subject here.

*The integrability conditions of these equations must of course be fulfilled which limits the choice of the fields.

VII. APPROACHES TO THE LARGE NUMBER HYPOTHESIS

The method developed in the previous sections offers the possibility to construct models of the universe as the minimal invariant varieties of various groups of transformations and allows to introduce gauge fields and a general metric by breaking the symmetry of the group. This scheme is no doubt too general to describe nature.

We can require that the unperturbed space in which no localized masses are present be homogenous and expand with the increase of a cosmic time coordinate t. This suggests itself as a first step to modify the gravitational theory in order to make it compatible with Dirac's large number hypothesis.

We assume thus that the three dimensional space be of constant curvature at any value of t and its radius R expands as a function of t.

$$(ds)^2 = \frac{f^2(t)}{(1+\varepsilon\frac{r^2}{4})^2}\,\overset{\circ}{g}_{ik}\,dx^i dx^k - c^2(dt)^2 \qquad (1)$$

$$r^2 = \overset{\circ}{g}_{ik} x^i x^k$$

Where ε which determines the sign of the curvature may assume the values ± 1 or 0.

The function f(t) is restricted by Killings equations (II,2) to solutions of

$$ff'' - (f')^2 = \text{const.} \qquad (2)$$

if Killing vectors with nonvanishing time components should exist. The solutions of this equation contain either exponential functions of t as $f(t) = C\cosh C_1 t$ or they are periodic as $f(t) = C\cos C_2 t$ Neither of these solutions is compatible with Diracs large number hypothesis and observations: The periodic solution has to be excluded because it results in a maximal extension which in case of the universe is necessarily large in atomic units.

The nonperiodic case would be compatible with observations only if in atomic units C_1 is very small so that its reciprocal is again a large time independent dimensionless number which has to be excluded. There seems to be no way out then to abandon Killing vectors with t-components if one wants a theory which does not involve a large dimensionless number in cosmology. We must thus restrict the symmetry to that of a homogenous 3-space. Then we can consider the metric

$$(ds)^2 = \frac{t^2}{(1+\frac{\varepsilon r^2}{4})^2}\,\overset{\circ}{g}_{ik}\,dx^i dx^k - (c^2 dt^2) \qquad (1a)$$

of unperturbed space. We can still introduce gravitational fields due to localized masses by breaking the symmetry of space to obtain a general metric:

$$(ds)^2 = g_{ik}(\bar{x},t)\, dx^i dx^k - (c\,dt)^2 \qquad \text{(1b)}$$

in Gaussean coordinates.

The large number hypothesis rejects any time independent dimensionless large number and thus also a constant number of particles (e.g., fermions plus antifermions) in the universe. This change of particle number in Dirac's large number hypothesis is the point most difficult to perceive. Dirac has interpreted it sometimes as influx into a given volume, sometimes as true particle creation. Let me point out that within the general theory of relativity particle creation is not at all contradictory. The theory is not really known to deal with the particle character of matter but the energy of matter may change with time. In situations where the energy conservation law is understood, we know it may do so on the expense of gravitational energy and in other cases it may do so just as well. We know only of no detailed mechanism that produces matter at an adequate rate;* such a mechanism anyway is beyond the theory.

Let us look at the density of the Einstein tensor of the metric Eq. (1a):

$$\sqrt{g}\left(R_{44} - \frac{1}{2} g_{44} R\right)$$

is proportional to zero for $\varepsilon = -1$ and to t for $\varepsilon = 0$ and +1. Set this equal to $\kappa\sqrt{-g}\,T_{44}$ and assume that $\kappa :: t^{-1}$ thus $\sqrt{-g}\,T_{44} :: t^2$ and the same is true for an integral over all space. Our assumptions Eq. (1a) and $\kappa :: t^{-1}$ lead thus in the terminology of Dirac[5] to additive matter creation uniformly distributed over unperturbed space. κ has to be made here a dimensionless number, expressed in terms of suitable atomic units.

---

*The fact that quantum laws and time dependent gravitational fields imply the pair creation of matter was demonstrated and admirably worked out by Schrödinger as early as 1939 for cosmology.[19] It remained since then common knowledge with a few gravitational physicists dealing with elementary particle processes.[20] Schrödinger had already found that only critical phases of cosmology would lead to a significant rate of particle creation. L. Parker and later some other young physicists rediscovered the effect in the context of quantum field theory and Parker and his group performed a lot of useful work on the subject. For some relativists the effect then became a new revelation.

The choice f(t) = t and the Einstein tensor result in a pressure which is anomalous for matter at rest. I do not think that it must for this reason be given up because matter spontaneously created is not merely matter at rest. Moreover, the problem of the energy momentum density of the vacuum is not solved in gravitational theory. I have previously pointed out that quantum corrections lead to a small addition to the Einstein tensor in the gravitational field equations of terms with fourth order derivatives.[21] Later I concluded from invariance arguments of the coupling of the Dirac electron equation to electromagnetism and gravitation that the fourth order term with a dimensionless coupling constant by far preponderates.[22] The occurrence of fourth derivatives may appear as a drawback--but if the gauge potentials A_k discussed in the last section are used instead of the $g_{\mu\nu}$, the equations which unite gravity and electrodynamics in the Lagrangian $L = F_{\mu\nu}F^{\mu\nu}$ contain preponderately $R_{\mu\nu\rho\sigma}R^{\mu\nu\rho\sigma}$ and the Maxwell* Lagrangian, and the equations are of the second order.

Localized masses produce a perturbation of the metric. We assume the condition that at large distances from the perturbation the metric tends towards the form of eq. (la). For example, all the matter within a given domain may be concentrated in a small area at a given time t. The remaining space will however, not be matter free at later times. At this stage the way in which the cosmic time coordinate is continued into the domain of the perturbation is crucial for the theory. Dirac in his most recent paper obtains a modified spherically symmetric metric which may not result in a black hole. This result is still obtained by postulating that the measurements in atomic units are given by a conformal transformation from Einstein theory and by a special identification of t close to the central mass.

The generalization of the fundamentals introduced in the previous sections open a wider range of possibilities to the approach. I hope that in a final formulation of the field equations the gravitational constant will appear as a purely geometrical entity related to the boundary conditions of the expounding universe.

---

*The use of $R_{\mu\nu\rho\sigma}R^{\mu\nu\rho\sigma}$ and the Maxwell Lagrangian to unify electromagnetism and gravitation seems to have been taken up again recently by Zumino et al. in the context of supergravity but it does not contain the above features.

REFERENCES

1. P. Bergmann, Private Communication.
2. P. Jordan "Schwerkraft and Weltfall," Vieweg & Sohn, Braunschweig 1955.
3. C. Brans and R. Dicke, Phys. Rev. 124 p. 925 (1961).
4. P.A.M. Dirac, Proc. Roy. Soc. Lond. A 333 p. 403 (1973).
5. P.A.M. Dirac, Proc. Roy. Soc. Lond. A 338 p. 439 (1974).
6. P.A.M. Dirac, "A New Approach to Cosmology" Preprint Florida State University, 1978.
7. P.A.M. Dirac, Proc. Roy. Society A 165 p. 199 (1938).
8. "On the Measurement of Cosmological Variations of the Gravitational Constant", University Presses of Florida 1978 L. Halpern editor.
9. L. Halpern, General Relativity and Gravitation Vol. 8 Nr 8 p. 623 (1977).
10. S. W. MacDowell & F. Mansouri, Phys. Rev. Letters 38, p. 739 (1977).
11. L. P. Eisenhart, "Continuous Groups of Transformations" Princeton University Press 1933, Several reprints by Dover, especially §1, 3, 12, 21, and Chapter V.
12. L. P. Eisenhart, "Riemannian Geometry" Princeton Univ. Press (1964).
13. P.A.M. Dirac, Ann. Math, 30 p. 657 (1935).
14. P.A.M. Dirac, Ann. Math, ibid (1936).
15. See Ref. 11 §21.
16. See Ref. 11 §54.
17. F. Gürsey, Istambul Summer School of Theoretical Physics Gordon and Breach, New York 1962.
18. R. Utiyama, Physical Review 101, p. 1597 (1956).
19. E. Schrödinger, Physica 6 p. 899 (1939) and Proc. Roy. Irish Acad. XLVI A 25, (1940).
20. B. Dewitt, Thesis Harvard 1952, S. Gupta Proc. Phys. Soc. 65A p. 161 (1952) L. Halpern Nuovo Cim. X 33, p. 728 (1964).
21. L. Halpern, Ark. f. Fysik 34, p. 539 (1966).
22. L. Halpern in Lecture Notes in Mathematics 570 p. 355 Springer Berlin, Heidelberg, New York 1977 and FSU Preprints HEP 76-11-16 and HEP 75-12-30.

POINCARE INVARIANCE WITHOUT POINCARE GROUP

Siegfried A. Wouthuysen
Instituut voor Theoretische Fysica, Valckenierstraat 65,
1018 XE AMSTERDAM, The Netherlands

ABSTRACT

The existence of only one polarization for the massless spin $\frac{1}{2}$ particles in nature (neutrinos) is considered as an important feature of the observed lack of space-reflection invariance. This feature, although permitted for massless particles in the framework of the Poincaré group as external symmetry group, is not a necessary consequence of this framework.

A new group of external symmetry is proposed, which induces in space-time a structure more primitive than the usual metric structure. Supplementary assumptions, referring to the generation of a mass term for the spinor field, and to an internal supersymmetry between the spinor field and the photon fields, are formulated, which then lead to the usual equations of spinor-electrodynamics, and thus to the empirically confirmed Lorentz invariance of electrodynamics.

A parallel treatment of the free massless spinor field entails the suppression of one of the two polarization states, for particles as well as antiparticles, in the way observed in nature.

INTRODUCTION

Dirac's equation, and the Poincaré invariant formalism of quantum-electrodynamics which was built on it, provide a very satisfactory description of charged leptons and their interactions with the electromagnetic field, as far as these have been studied in detail at all.

Putting the mass parameter equal to zero, Dirac's equation yields a theoretical frame which contains more than the states actually found in nature for the massless leptons (neutrinos). The necessary restriction can be put in ad hoc, and is compatible with the continuous Poincaré group. However, from the point of view of the complete external symmetry group, no reason is apparent which would restrict the solutions of the massless Dirac equation to the ones observed.

The work reported here is aimed at finding a frame for the description of spin $\frac{1}{2}$ particles which will automatically lead to the desired difference between massive and massless particles. A clue can be found by studying the symmetry properties of limiting cases of Dirac's equation. As is well known, the ordinary mass zero equation not only allows the Poincaré group: it is even invariant under the conformal group. However, also equations of the form:

$$\not{p}\psi \;=\; \alpha P_{\pm}\psi \quad (\text{with} \quad P_{\pm} \;=\; \tfrac{1}{2}(1 \pm \gamma^5)\ )$$

ISSN: 0094-243X/78/153/$1.50

can be considered as mass zero limits of Dirac's equation[1]. Their solutions allow a symmetry group - leaving invariant a nonsingular bilinear in ψ and $\psi^\dagger$ - which again differs from the Poincaré group. This group G, the same for both branches (with righthand members αP_+ or αP_-), has as generators of its continuous part space translations and rotations, time translation and space-time dilatation. As discrete element we take it to contain spatial inversion. Its Lie algebra is only eight dimensional, smaller than the ten dimensional one of the Poincaré group. Like the Poincaré group it is a subgroup of the causal group[2]. It induces in space-time a structure which is more primitive than the metrical Poincaré structure. G is the group that transforms pairs of neighboring points in space-time in such a way that the absolute value of the velocity leading from one to the other is invariant, while their order in time is maintained. Note, that G does not contain transformations to moving coordinate systems (boosts).

It is this group G, discovered in a heuristic way as the symmetry group of certain limiting cases of Dirac's equation, which we shall from now on adopt as the external (space-time) group.

SPINOR FIELDS IN G SPACE-TIME

If one assumes space-time to be structured according to the group G, one is led to consider unitary representations of G, as a preparation to the establishing of field equations for free particles[3,4]. If the generators for translations in space and time, the space rotations and the dilatation are respectively written as $\vec{P}$, H, $\vec{J}$, D, it is straightforward to establish $c = H|P|^{-1}$ and $\eta = (\vec{P}.\vec{J})|P|^{-1}$ as the two Casimir operators of the continuous part of G. If one includes spatial inversion, the two Casimir operators will be c and $|\eta|$. Here, c can have any real value, and $|\eta|$ can be 0, $\frac{1}{2}$, 1,.... In what follows, we shall restrict ourselves, in view of our purpose, to the case $|\eta| = \frac{1}{2}$. The value of the velocity c determines the opening of the cone in four-momentum space, which describes the dispersion law proper to the representation.

The irreducible representations can be implemented by fields, obeying G-invariant differential equations, the solutions of which transform according to a reducible representation containing two irreducible ones with opposite values of c, and the same value of $|\eta|$, in a way similar to what happens in the Poincaré case.

The invariant norm is given by the spatial integral of a local density, built bilinearly from a solution and its complex conjugate. For the spinor field this G-invariant norm (or charge) is uniquely determined, and it is indefinite: the wave equation is simply d'Alemberts equation (with the characteristic velocity c) for a two-component spinor field, and the charge is the indefinite Klein-Gordon charge, having a sign changing with that of the energy. This fact prevents quantization of the G-spinor field according to the exclusion principle.

MASS AND THE NEW CHARGE DENSITIES

We now temporarily abandon the discussion of spinor fields invariant under the complete group G; as indicated above, this invariance leads to a conical dispersion law, characteristic - in the language of relativity - for massless particles. The concept of mass is, of course, foreign to the group G, but we shall now assume that the dilatation invariance of the spinor field equation is broken by a "mass-term", that is a term proportional to the field such that the field equation still has two disjoint sets of solutions, with positive and negative frequencies, respectively. As the G-covariance now is broken, the requirements on the charge become less stringent, and, as it turns out, an infinite (denumerable) set of possible charge densities exist, invariant under a truncated G-group[5,6]. This truncated group not only lacks the dilatation, but also the space-inversion. The choice of charge density is closely linked to that of a local antilinear involution among the solutions of the wave equation, corresponding to charge conjugation[7]. The possible new charge conjugations are numbered by a rank number n, assuming all positive and negative integer values. The value zero leads back to the old G-invariant charge density, with an indefinite charge. Also, all even values of n lead to indefinite charges. However, all odd values of n lead to positive definite charges, and therefore, to a theory amenable to quantization according to the exclusion principle. If therefore, one requires Fermi-Dirac statistics, one is forced to take an odd value of n. The simplest choices for n turn out to be ± 1 (discussion of other odd values of n is postponed to a forthcoming paper). Both of these lead to a field equivalent to a Dirac field; and in both these cases the equations and the charge are also invariant under suitably defined boosts, and under a suitably defined spatial inversion, completing the truncated G-group to the orthochronous Poincaré group.

A G-covariant spinor field leads therefore, after breaking the dilatation invariance by a mass-term, by the mere requirement of a positive definite charge (and a simplicity argument in achieving this) to the Lorentz invariant field-equation for free massive spin $\frac{1}{2}$ particles. Note, that the relation between spin and statistics for this special case (spin $\frac{1}{2}$, Fermi-statistics) is used here as an input, which contributes to the Lorentz invariance of the ensuing field equation. The G-structured space-time does not in itself lead to a relation between spin and statistics.

INTERACTION WITH THE ELECTROMAGNETIC FIELD

The considerations of the preceding section allow straightforward generalization to the case where an external electromagnetic field is present. Along exactly the same lines the choices ± 1 for n lead to Dirac's equation in an external field with minimal coupling.

The situation changes if one wants to include the free electromagnetic modes in the formalism. The G-structure of space-time suggests description by a scalar and a 3-vector field, in accordance with the idea of gauge transformation of the spinor field: the gauge-invariant derivatives of the spinor field require a scalar for the time-derivative and a vector for the spatial derivatives. The corresponding free fields would, however, in general correspond to different values of the Casimir operator c. The description of the spinor field in interaction with the scalar-vector electromagnetic field would, in general, involve three characteristic velocities c, c_s and c_V. The empirical evidence forces us, clearly, to believe in a degeneracy of these Casimir operators for the three fields. Once this degeneracy is put in - we come back to it in the following section - we obtain the usual equations of spinor-electrodynamics, with its Lorentz invariance as a "bonus".

INTERNAL SUPERSYMMETRY

In the preceding sections an important role was played by two supplementary inputs: the imposed statistics, both for the G-spinor and for the G-scalar- and vector fields, and the imposed degeneracy of the characteristic velocities. In this section, we indicate how, due to the structure of the G-group, these two inputs can be concisely combined in a single assumption of internal symmetry. Indeed, an internal symmetry is the traditional way to account for a degeneracy outside the scope of the external symmetry group, a degeneracy of external characteristics. The equality of the Casimir operators c of the scalar, spinor and vector fields would require, in view of the statistics to be imposed, an internal supersymmetry or Fermi-Bose symmetry[8]. The field components, two complex and four real ones in any space-time point, should form a supermultiplet of this supersymmetry. As graded Lie algebra generating the suppersymmetry we take $SU(2|1)$; its even components form the Lie algebra $SU(2) \otimes U(1)$. And we assume the field components to form a basis for its regular representation. Under the SU(2) internal subgroup, the representation reduces, the irreducible parts being a scalar, a spinor (and its conjugate) and a (three) vector. The U(1) internal subgroup is effectively the gauge group.

It is to be remarked that the SU(2) internal subgroup plays the role of the spin part of the G-rotation: the spin appears at this stage as an internal degree of freedom, while the external group G only governs the orbital motion. The relevant part of the symmetry group is therefore $G \otimes SU(2)$. Only in the next stage, where a positive definite charge has to be defined for the fermions (after introduction of the mass term), it becomes essential that spin and orbit transform together. The pertinent subgroup of $G \otimes SU(2)$ is then again G, due to the structure of G. A gauge-invariant theory involving only the fields thus obtained leads then necessarily to the equations of spinor-electrodynamics, with again as a bonus the full Lorentz invariance.

HANDEDNESS OF THE NEUTRINO

The above discussions indicated that the same Dirac field can be obtained from a G-covariant two-component spinor field, after breaking dilatation invariance by a mass term, in two ways, depending on the choice ± 1 for n in constructing the charge density. The original two-component field appears either as the $\gamma^5 = +1$ or as the $\gamma^5 = -1$ part of the Dirac field.

If one now keeps one of the charge (or norm) definitions just mentioned, but makes the mass tend to zero, one obtains in the limit a (two-component) G-covariant field with a norm that is still only invariant under the broken group. This norm is in the limit however semi-definite: particle states of one helicity and antiparticle states of the opposite helicity have norm zero, while the remaining states all have positive norm. Again, the invariance group can be extended to comprise the full Poincaré group. This we consider as the possible reason why the neutrinos in nature lack half of their degrees of freedom. If this connection were borne out, we have to consider the breaking of parity in weak interactions as an - albeit indirect - indication for the relevance of a preferred rest-frame in particle physics.

CONCLUSION AND OUTLOOK

The work that has been sketched, in broad outline, in the present paper, is partly already in the literature in a preliminary form; a more systematic version is in the process of being written.

The only result these rather incisive changes in the basic framework of particle theory seem to bring forth, is an indication for the reason of the breaking of parity in leptonic weak interactions. The preferred rest-frame remains masked in the applications we discussed: both spinor electrodynamics and the semi-definite norm of the neutrino field are invariant, at least under continuous Lorentz transformations. However, this masking effect arises directly from the simplifying choices ± 1 for the parameter n. Other odd values of n seem to lead to wave equations for massive spin $\frac{1}{2}$ particles, where the orbital motion, even in an external electromagnetic field, follows exactly that of Dirac particles. The only domain where the original preferred rest-frame may manifest itself is that of non-electromagnetic effects of the spin; but this has as yet not been investigated in detail. Also, experimental evidence seems to be lacking here.

In particular, it would be interesting to study closely strong interaction processes in proton-proton collisions, both with a stationary target configuration and with colliding beams, and compare results in domains of similar center of mass energies.

We refrain from commenting on the possible implications of a more philosophical nature, apart from stating that this development might appeal to advocates of Mach's principle in one form or another: fewer properties of empty space, more assumptions about relations between different manifestations of matter, enter our considerations, than in the usual framework.

REFERENCES

1. S. A. Wouthuysen, Proc. Kon. Ned. Ak. v. Wet. 168, 54 (1958).
2. E. C. Zeeman, J. Math. Phys. 5, 490 (1964).
3. S. Weinberg, Phys. Rev. B133, 1318 (1964).
   Phys. Rev. B134, 882 (1964).
4. E. A. de Kerf, Thesis, University of Amsterdam, 1967.
5. E. Mendels, Proc. Kon. Ned. Akad. Amsterdam, B76, 50, 61, 71, 86, 105, 122 (1973).
6. E. Mendels and S. A. Wouthuysen, Phys. Lett. 43B, 333 (1973).
7. S. A. Wouthuysen, Spinor fields in a space-time with submetrical structure (to appear in Physica).
8. L. H. Karsten, Thesis, University of Amsterdam (in preparation).

ELIMINATION OF REDUNDANCY IN SUPERSPACE EQUATIONS OF SUPERGRAVITY[+*]

Murray Gell-Mann, Pierre Ramond[**], and John H. Schwarz[***]

California Institute of Technology, Pasadena, California 91125, U.S.A.

ABSTRACT

The equations of N=1 supergravity in superspace, with $e\neq 0$ and with external vector-spinor matter, were derived previously, but in redundant form (420 equations for 112 potentials V_Λ^A and h_Λ^{rs}.) There are many possible "minimal lists" of 112 equations from which the other equations can be obtained by means of Bianchi identities for the field strengths. Here we exhibit such a minimal list and derive the full set of equations. The procedure can easily be generalized to extended supergravity (N>1).

- - - - - - - - - - - - - -

Equations of motion in superspace have been given for N=1 supergravity,[1,2] including $e\neq 0$[2] and vector-spinor matter, and for N=3 supergravity including $e\neq 0$[3]. In this communication we restrict ourselves to the case N=1 and employ the notation of Ref. 2.

The equations of motion are given in terms of curvatures R_{AB}^{rs} and torsions S_{AB}^{C}. These are defined in terms of the vielbein V_Λ^A with 64 components and the connection h_Λ^{rs} with 48 components. Hence we expect 112 independent equations of motion. However, 420 such equations were given, and it is important to eliminate the redundancy and exhibit 112 equations from which the rest can be derived using Bianchi identities. We need to employ only the following identities connecting Latin covariant derivatives $\mathcal{D}_A$ in superspace:

$$[\{\mathcal{D}_a,\mathcal{D}_b\},\mathcal{D}_c] + [\{\mathcal{D}_b,\mathcal{D}_c\},\mathcal{D}_a] + [\{\mathcal{D}_c,\mathcal{D}_a\},\mathcal{D}_b] = 0, \quad (1)$$

$$[\{\mathcal{D}_a,\mathcal{D}_b\},\mathcal{D}_r] = \{[\mathcal{D}_a,\mathcal{D}_r],\mathcal{D}_b\} + \{[\mathcal{D}_b,\mathcal{D}_r],\mathcal{D}_a\}, \quad (2)$$

out of the set of Latin Bianchi identities

$$(-1)^{AC}[[\mathcal{D}_A,\mathcal{D}_B\},\mathcal{D}_C\} + \text{cyclic permutations} = 0.$$

+ The review talk by M. Gell-Mann at the Symposium honoring Professor Dirac in May, 1978 included a preliminary report on this research.

* Work supported in part by the U.S. Department of Energy under Contract No. EY76-C-03-0068 and by the Alfred P. Sloan Foundation.

** Robert Andrews Millikan Senior Research Fellow.

*** John Simon Guggenheim Fellow.

ISSN: 0094-243X/78/159/$1.50

-2-

We have, of course,

$$[\mathcal{D}_A, \mathcal{D}_B\} = R_{AB}^{C}\,\mathcal{D}_C + \frac{1}{2}\,R_{AB}^{rs}X_{rs} \equiv R_{AB}^{Y}G_Y, \tag{3}$$

where we define

$$R_{AB}^{C} = -2S_{AB}^{C}. \tag{4}$$

We begin with the set of 40 equations

$$R_{ab}^{r} = i(\gamma^r)_{ab}, \qquad \text{(40 eqs.)} \tag{5}$$

which obtain even in the presence of spinor-scalar matter. Utilizing the coefficient of $\mathcal{D}_t$ in Eq. (1), we find that R_{ab}^{c} can be expressed in terms of quantities P_b and T_b^{st} (antisymmetric in s and t):

$$R_{ab}^{c} = i(\gamma^r)_{ab}(2\gamma_r\gamma_5 P - \tfrac{i}{2}[\sigma_{st},\gamma_r]T^{st})^c + 1/2(\sigma^{ru})_{ab} \text{ times}$$

$$([\sigma_{ru},\sigma_{st}]T^{st} + 8i\sigma_{ru}\gamma_5 P)^c \tag{6}$$

and that R_{ra}^{t} can be expressed in terms of T and a quantity B_b^t:

$$R_{ra}^{t} = iB_b^t(\gamma_r)^b{}_a - \frac{1}{2}(T^{uv}[\gamma_r,\sigma_{uv}]\gamma^t)_a + \frac{1}{4}(T^{uv}[\sigma_{uv},\sigma_{rs}]\gamma^t\gamma^s)_a. \tag{7}$$

Since R_{ra}^{t} vanishes for supergravity, we have the equations

$$B_b^t = 0, \qquad \text{(16 eqs.)} \tag{8}$$

$$T_b^{st} = 0. \qquad \text{(24 eqs.)} \tag{9}$$

When spinor-scalar matter is introduced, R_{ab}^{c} and P_c are non-zero, but from now on we assume such matter is absent, so that $R_{ab}^{c} = 0$ and hence

$$P_b = 0. \qquad \text{(4 eqs.)} \tag{10}$$

We now set the coefficient of $\mathcal{D}_{rs}$ in Eq. (1) equal to zero and learn that if we define f_t^{rs} and f_{tu} by means of the relation

$$R_{ab}^{rs} = i(\gamma^t)_{ab}\, f_t^{rs} + (\sigma^{tu})_{ab}\, f_{tu}^{rs}, \tag{11}$$

then

$$f_{tu}^{rs} = C_t^r\delta_u^s - C_t^s\delta_u^r - C_u^r\delta_t^s + C_u^s\delta_t^r + \varepsilon_t{}^{rsv}D_{uv} - \varepsilon_u{}^{rsv}D_{tv}, \tag{12}$$

where $C_{rt} = C_{tr}$ and $D_{uv} = D_{vu}$, and

$$R_{rc}^{d} = iB_e^d(\gamma_r)^e{}_c - \frac{1}{2}f_r^{tu}(\sigma_{tu})^d{}_c - \frac{i}{2}C_{rt}(\gamma^t)^d{}_c - \frac{1}{2}D_{rt}(\gamma_5\gamma^t)^d{}_c. \tag{13}$$

The equations of motion tell us that

$$f_t^{rs} = 0. \qquad \text{(24 eqs.)} \qquad (14)$$

Next, we set the coefficient of X_{rs} in Eq. (1) equal to zero and learn that

$$R_{ra}^{tu} = i(B^{tu}\gamma_r)_a \qquad (15)$$

and that the tensors C_{rt} and D_{rt} are covariant constants, which means they are equal to constants times η_{rt}. Such a term in the case of D_{rt} would violate parity conservation, and it can be rotated away in any case by a chiral transformation of the superspace spinor coordinate θ, so we put $D_{rt} = 0$. Then, if we define the integration constant e by means of the relation

$$(\sigma_{rs})^{ab}R_{ab}^{rs} = -24e, \qquad (16)$$

we obtain the results

$$R_{ab}^{rs} = -2e(\sigma^{rs})_{ab}, \qquad (17)$$

$$R_{rc}^{d} = iB_e^d(\gamma_r)^e{}_c + i\,\frac{e}{4}(\gamma_r)^d{}_c. \qquad (18)$$

Setting the coefficient of $\mathcal{D}_t$ in Eq. (2) equal to zero, we obtain

$$B_c^d = \frac{e}{4}\,\delta_c^d - (\gamma_5\gamma_t)^d{}_c J_5^t \qquad (19)$$

and

$$R_{rs}^t = 2\varepsilon^t{}_{rsu}J_5^u, \qquad (20)$$

where J_5^t is so far unrestricted.

The vanishing of the coefficient of $\mathcal{D}_d$ in Eq. (2) tells us that we have

$$\mathcal{D}_a J_5^r = \frac{1}{2}(\gamma_5\gamma_t)^b{}_a B_b^{rt}, \qquad (21)$$

$$(\gamma_r)^{ab}\mathcal{D}_a J_5^r = 0, \qquad (22)$$

$$R_{rs}^d = -\eta_{rt}\,\eta_{su}(1)^{da}B_a^{tu}. \qquad (23)$$

Comparison with the equations of motion of Ref. 2 shows that J_5^r is just the axial vector superfield describing vector-spinor matter and obeying precisely the constraint Eq. (22), which implies, among other things, that $\mathcal{D}_r J_5^r = 0$. The specification of J_5^r

$$J_5^r = J_5^r \text{ (vector-spinor matter)} \qquad \text{(4 eqs.)} \qquad (24)$$

thus completes our list of 112 independent equations of motion.

The remaining equation of Ref. 2 is now easily derived. We examine the coefficient of X_{rs} in Eq. (2) and obtain

-4-

$$R^{tu}_{rs} = -(\sigma_{rs})^{bc}\mathcal{D}_b B^{tu}_c + e^2(\delta^t_r\delta^u_s - \delta^t_s\delta^u_r). \tag{25}$$

Using Eq. (21), we obtain in addition the result

$$R^{tr}_{rs} = -(\gamma_5\gamma_s)^{ab}\mathcal{D}_a\mathcal{D}_b J^t_5 - 3e^2\delta^t_s, \tag{26}$$

the Einstein equation in superspace.

To summarize, the superspace equations of motion of Ref. 2 have now been reduced to the non-redundant set of Eqs. (5, 8, 9, 10, 14, 24) and the specification in Eq. (16) of the integration constant e. All of these relations can be expressed in terms of torsion components.

As this research was being completed, we were told by Samuel MacDowell of his work on the same subject[4] (including higher N, but without external matter.) It is a pleasure to acknowledge useful and enlightening conversations and correspondence with him. We would like to acknowledge also the hospitality of the Aspen Center for Physics, where this manuscript was prepared.

REFERENCES

1. J. Wess and B. Zumino, Phys. Letters 66B, 361 (1977).
2. L. Brink, M. Gell-Mann, P. Ramond, and J.H. Schwarz, Phys. Letters 74B, 336 (1978); erratum to be published.
3. L. Brink, M. Gell-Mann, P. Ramond, and J.H. Schwarz, Phys. Letters 76B, 417 (1978).
4. S. MacDowell, Yale preprint. We are indebted to Professor MacDowell for communicating his results to us prior to publication.

VARIATION OF CONSTANTS

Freeman J. Dyson
Institute for Advanced Study, Princeton, NJ, 08540.

ABSTRACT

A summary of recent observational evidence bearing on the question of the possible variation of the natural constants with time. No conclusive evidence of variation is found.

1. INTRODUCTION

Six years ago I wrote a review article[1] summarizing the evidence for and against Dirac's hypothesis[2] that the large dimensionless numbers appearing in the laws of physics are increasing with time. I described five alternative hypotheses defining the way the constants might be varying, and ended my review with the words: "It is quite possible that all five will fail, and then it will be up to the speculative cosmologists, and up to Dirac in particular, to think of something new." Now, six years later, although two of the five hypotheses are still viable, Dirac has thought of something new.[3] I am sorry that I have not had time to study Dirac's new theory[3] in detail, nor the very interesting papers of Canuto and his collaborators[4] in which Dirac's cosmology is analyzed in depth. I am sorry that Canuto is not here to comment on the new ideas of Dirac that we shall hear this morning.[5] All that I shall do in this talk is to bring my 1972 review[1] up-to-date by describing the new observational evidence that has appeared in the last six years concerning the possible variation of the constants. The only strong conclusion that my 1972 article contained was the upper limit

$$\left| \frac{1}{\alpha}\frac{d\alpha}{dt} \right| < 5.10^{-15}\ \text{year}^{-1}\ , \tag{1}$$

on the variation of the fine-structure constant α, obtained from an analysis of the abundance ratios of rhenium and osmium isotopes in iron meteorites and molybdenite ores. The reason why these isotope-ratios are a sensitive indicator of variation of α is[6] that the decay of Rhenium 187 to Osmium 187 has a long half-life and the exceptionally small decay-energy of 2.5 Kev. The effect of a change in α on the decay is amplified by the ratio of the Coulomb energy of a proton in the nucleus to the decay-energy, in this case by a factor of about 20000.

2. SHLYAKHTER

A spectacular improvement on the upper limit (1) for variation of α was obtained by Shlyakhter[7] from an analysis of isotope ratios in the natural fission reactor[8] that operated about 2.10^9 years ago in the ore body of the Oklo uranium mine in Gabon, West Africa.

ISSN: 0094-243X/78/163/$1.50

The crucial quantity is the ratio (Sm 149/Sm 147) between the abundances of two light isotopes of samarium which are not fission products. In normal samarium this ratio is 0.9, in the Oklo reactor it is about 0.02. Evidently the Samarium 149 has been heavily depleted by the dose of thermal neutrons to which it was exposed during the operation of the reactor. The fluence (integrated dose) of neutrons can be calculated from the measured ratios of (U 235/U 238) and (Nd 143/Nd 142) in the same ore samples, and is found to be about $1.5 \cdot 10^{21}$ neutrons cm^{-2}. A detailed analysis of the data gives the result

$$\sigma = 55 \pm 8 \text{ Kilobarn} \tag{2}$$

for the capture cross-section of thermal neutrons by Samarium 149 two billion years ago. This agrees with the modern value. Now the cross-section (2) is dominated by a capture resonance at a neutron energy of 0.1 ev. If the energy difference between the ground-state of Sm 149 and the compound nucleus Sm 150 had varied by as much as 0.02 ev between Oklo time and the present, the cross-section (2) would be more than two standard deviations away from its present value. This neutron resonance provides a far more sensitive test for variation of constants than the Re 187 decay. Roughly speaking, the effect of a variation of α will be amplified in the neutron cross-section by a factor 10^8, the ratio of the neutron binding energy to the neutron resonance energy in Sm 150. Shlyakhter, using some detailed assumptions about the nuclear physics which I will not discuss here, deduces from (2) the upper limit

$$\left| \frac{1}{\alpha} \frac{d\alpha}{dt} \right| < 5.10^{-18} \text{ year}^{-1} . \tag{3}$$

I have checked his numbers and find that (3) is a conservative estimate. Anyone with access to the original Oklo data could probably push the upper limit down even further.

Another important by-product of Shlyakhter's analysis is an upper limit

$$\left| \frac{1}{\beta} \frac{d\beta}{dt} \right| < 10^{-12} \text{ year}^{-1} \tag{4}$$

on the time-variation of the dimensionless ratio

$$\beta = (gm^2 c/\hbar^3) = 9.10^{-6} \tag{5}$$

which measures the strength of the weak-interaction coupling constant g. The estimate (4) is obtained from the same data as (3), assuming that weak interactions contribute a fraction of the order of 10^{-7} to nuclear binding. From (4) we can already exclude with some degree of certainty the hypothesis that β might be varying with cosmic time t with some small negative exponent such as $t^{-(1/8)}$.

3. WOLFE, BROWN AND ROBERTS

Wolfe, Brown and Roberts[9] measured the red-shift

$$Z_H = 0.52385 \pm 0.00001 \qquad (6)$$

in the absorption by neutral hydrogen of radio waves from the BL-Lac object AO 0235+164. They pointed out that the optical red-shift of the same object,

$$Z_{Mg} = 0.52392 \pm 0.0001 , \qquad (7)$$

deduced from the absorption of light by magnesium ions, agrees with (6) within the accuracy of the measurements. The frequency of the radio-absorption depends on the proton magnetic moment, whereas the optical frequency depends only on the properties of the electron. The ratio of the two frequencies is therefore proportional to the quantity

$$P = \alpha^2 g_P(m_e/m_P) , \qquad (8)$$

and the equality of the red-shifts implies an upper bound

$$\left| \frac{1}{P}\frac{dP}{dt} \right| < 2.10^{-14} \text{ year}^{-1} . \qquad (9)$$

Here g_P is the proton gyromagnetic ratio, m_e and m_P the electron and proton masses. The bound (9) thus implies an equally strong bound on possible variation with time of the electron-proton mass-ratio.

4. REASENBERG AND SHAPIRO

We now come to the question of the variation of the gravitational constant G, or of the dimensionless ratio

$$\gamma = Gm^2/\hbar c = 5.10^{-39} , \qquad (10)$$

where m is the proton mass. The central feature of Dirac's cosmology is that G and γ should vary as t^{-1}. Reasenberg and Shapiro[10] have observed the planets Mercury, Venus and Mars with radar. If G were decreasing, there should be a secular increase in the radii and in the periods of the planetary orbits. Unfortunately the effects of varying G can only be disentangled from the effects of mutual planetary perturbations by a very precise and elaborate integration of the equations of motion of the entire solar system extending over many years of data. The data from the period 1966 to 1975 lead to the conclusion

$$-0.5.10^{-10} \text{ year}^{-1} < \frac{1}{G}\frac{dG}{dt} < 1.5.10^{-10} \text{ year}^{-1} . \qquad (11)$$

This result is consistent with G varying like t^{-1} or with G staying constant. There are no other astronomical data comparable in accuracy and reliability with these radar observations.

Looking to the future, Reasenberg and Shapiro remark that the statistical uncertainty in (dG/dt) resulting from observations extending over T years is proportional to $T^{-5/2}$ if the precision of individual measurements remains constant. If the techniques of measurement improve, the uncertainty in (dG/dt) decreases with T even faster. Reasenberg and Shapiro are therefore confident that, if they are fortunate enough to continue their observations until 1985, they will then be able to determine (G^{-1} dG/dt) to a precision of one part in 10^{11} years. It will then be possible to distinguish clearly between the Dirac cosmology and the orthodox constant-G cosmology.

There is unfortunately an additional theoretical ambiguity in the analysis of the Reasenberg-Shapiro data. Reasenberg and Shapiro assumed in their analysis that the radius R and period P of each orbit vary with G according to the adiabatic law

$$R \sim G^{-1}, \; P \sim G^{-2} . \tag{12}$$

But Dirac's 1978 theory[3] breaks the adiabatic invariance and gives a different rule of evolution of the orbits,

$$R \sim G^{-1/3}, \; P \sim G^{-1} . \tag{13}$$

If the radar observations are to be used to test the Dirac cosmology, the entire analysis of the data must be carried through consistently, with the orbits varying according to (13). It is not possible in any simple way to correct the result (11), which Reasenberg and Shapiro calculated using the assumption (12), to find the limit on variation of G which would follow from the assumption (13). It is to be hoped that when Reasenberg and Shapiro analyze their data in 1985 they will consider both possibilities (12) and (13) and clear up this ambiguity once and for all.

5. VAN FLANDERN

Van Flandern[11] claims to have found positive evidence for a variation of G by analyzing observations of the motion of the moon. He discusses a quantity called the "lunar acceleration" which means the second derivative of the observed deviation of the moon from its calculated orbit. He looks at two series of observations, modern and old.

a) Modern measurements, made by timing with atomic clocks the occultation of stars by the edge of the moon. These extend over the time-base 1955-1974 and give for the acceleration the value

$$A = -65 \pm 18 \text{ arc sec/century}^2 . \tag{14}$$

b) Old measurements, made by observing the moon's position with clocks keeping "ephemeris time," i.e. with time defined by the motion of the earth around the sun. These measurements are less accurate but extend over a longer time-base, 1750-1970. They give the acceleration

$$A = -38 \pm 4 \text{ arc sec/century}^2 . \tag{15}$$

Now Van Flandern observes that the acceleration (14) should include the deviation of ephemeris time from atomic time produced by a variation of G, whereas the acceleration (15) refers only to ephemeris time and should be independent of any variation of G. So he claims that the difference between (14) and (15) gives a direct measurement of the variation of G, namely

$$\frac{1}{G}\frac{dG}{dt} = -7.5 \pm 2.7.10^{-11} \text{ year}^{-1} . \tag{16}$$

This rate of variation has the right sign and the right order of magnitude to agree with Dirac's cosmology.

Unfortunately the moon itself is not a good enough clock to make the difference between (14) and (15) significant. The acceleration (15) is averaged over 200 years while (14) is averaged over 20. There is no reason to believe that A remains constant to the required accuracy over 200 years. The tidal interactions of the moon with the earth can neither be measured nor calculated accurately enough to allow the effect on A of varying G to be isolated.

So I conclude that Van Flandern's case for a variation of G with time remains unproven.

REFERENCES

1. F. J. Dyson, "The Fundamental Constants and their Time Variation," in "Aspects of Quantum Theory," Ed. A. Salam and E. P. Wigner, (Cambridge U. Press, 1972) pp. 213-236.
2. P.A.M. Dirac, Nature, 139, 323 (1937), Proc. Roy. Soc. A165, 199 (1938).
3. P.A.M. Dirac, "A New Approach to Cosmological Theory," Florida State University preprint, (January 1978).
4. V. Canuto and J. Lodenquai, "Dirac Cosmology," and V. Canuto, "Observational Tests of Dirac's Cosmology," Institute for Space Studies preprints, (1978).
5. P.A.M. Dirac, these Proceedings, following lecture.
6. P. J. Peebles and R. H. Dicke, Phys. Rev. 128, 2006 (1962).
7. A. I. Shlyakhter, Nature, 264, 340 (1976), and Leningrad Nuclear Physics Institute preprint (1976). Unfortunately the version published in Nature omits most of the detailed analysis that appears in the preprint.
8. M. Maurette, "Fossil Nuclear Reactors," Ann. Rev. Nucl. Sci. 26, 319 (1976).
9. A. M. Wolfe, R. L. Brown and M. S. Roberts, Phys. Rev. Letters, 37, 179 (1976).
10. R. D. Reasenberg and I. I. Shapiro, "Bound on the Secular Variation of the Gravitational Interaction," MIT preprint (1975).
11. T. C. Van Flandern, Month. Not. Roy. Ast. Soc. 170, 333 (1975); Sci. Am. 234, 44 (Feb. 1976).

CONSEQUENCES OF VARYING G

P.A.M. Dirac
Florida State University
Tallahassee, Florida 32306

THE VARIATION OF G

What is the reason for believing that G varies? It comes simply from a consideration of the dimensionless fundamental constants that are provided by Nature. Most of these are fairly close to unity. There is $\hbar c/e^2$, which is approximately 137, and there is the ratio of the mass of the proton to that of the electron, m_p/m_e, which is approximately 1840. Another fundamental dimensionless constant is the ratio of the electric to the gravitational force between proton and electron. This has the approximate value.

$$\frac{e^2}{G\, m_p m_e} = 7 \times 10^{39} \tag{1}$$

Physicists believe that ultimately an explanation will be found for all these numbers. Those close to unity will be explained by mathematical theories. But how can a mathematical theory ever lead to a number of order 10^{39}? It seems hopeless.

It would be more reasonable to connect this number with another large number provided by the age of the Universe. The time since the Big Bang, the epoch, is around 18 x 10^9 years, if one accepts the latest estimate of the Hubble constant and the usual explanation for it. If we express it in terms of a unit of time provided by atomic constants, say $e^2/m_e c^3$, we get

$$t = 18 \times 10^9 \text{ years} = 2 \times 10^{39}\, e^2/m_e c^3 . \tag{2}$$

We get again a number of order 10^{39}.

Now one might consider this to be a remarkable coincidence. But it would be more reasonable to believe that there is some explanation for it, which will be understood when we know more about cosmology and atomic theory. The numbers (1) and (2) are then connected, with a coefficient close to unity. Since (2) increases with the epoch t, (1) must increase in the same ratio. We are thus led to

$$G :: t^{-1}$$

where G is the gravitational constant expressed in atomic units.

ISSN: 0094-243X/78/169/$1.50

If the foregoing argument is valid, it implies that all the large fundamental dimensionless constants provided by nature are varying, and are connected with t by simple equations with coefficients close to unity. This general conclusion we call the Large Numbers Hypothesis (L.N.H.).

An immediate consequence is that the Universe must continue to expand forever, and cannot attain a maximum size and then contract again, since the maximum size, expressed in atomic units, would give a large number not varying with t. Thus some popular models for the Universe are ruled out.

THE LAW OF RECESSION OF THE GALAXIES

Let us bring in another large number, provided by the total mass of the observable universe. To make it precise, let us take the total mass of that part of the universe that is receding from us with a velocity $< \frac{1}{2}c$. The fraction $\frac{1}{2}$ here is not important, since we are concerned only with orders of magnitude. To get a dimensionless number we express it in terms of the proton mass, and call it then N.

The value of N is not known very well, because one does not know how much invisible mass there is, but it is probably of the order 10^{78}. From the L.N.H. we then infer that

$$N :: t^2 . \qquad (3)$$

There is thus a continual increase in the amount of observable matter. How can one account for it? One might assume continuous creation of matter, so that the usual law of conservation of mass is only an approximation. I have been working with this assumption for a number of years, but find difficulties in reconciling it with various observations, and now believe it should be given up.

There is an alternative explanation which keeps to strict conservation of mass. We assume that the speed of recession of the galaxies is not constant, but they are steadily decelerating, so that more and more galaxies are continually appearing with velocity of recession $<\frac{1}{2}c$. This was the picture proposed in my original paper[1] on the subject.

We can easily work out what the law of deceleration must be to lead to the result (3). Let R be the distance, in atomic units, of galaxies with $V_{rec} = \frac{1}{2}c$. R is a measure of the radius of the observable universe. We have $R = ct$, with some numerical coefficient close to unity, which does not matter in the present discussion. Then the average density of matter, in atomic units, is

$$N/R^3 = N/(ct)^3 :: t^{-1}.$$

So the distance of each piece of matter $:: t^{1/3}$ and thus the distance of a galaxy is

$$D :: t^{1/3} \qquad (4)$$

The velocity of recession of a galaxy is now

$$v_{rec} :: t^{-2/3} \qquad (5)$$

There is thus a strong deceleration of the galaxies and a considerable departure from the usual picture of galaxies receding with almost constant velocity.

THE NATURAL MICROWAVE RADIATION

Is there any confirmation for this law of deceleration? There is very good confirmation, provided by the observation of the natural microwave radiation.

The microwave radiation is believed to be of cosmological origin because of its uniformity and isotropy. So far as it can be observed it appears to be black-body radiation, satisfying Planck's law. Now black-body radiation in an expanding universe remains black-body radiation (provided there is no creation of photons, such as might occur in a theory with continuous creation of matter). Each spectral component of the radiation gets red-shifted according to the same law as the distance of a galaxy, thus from (4)

$$\lambda :: t^{1/3} . \qquad (6)$$

The temperature T of the radiation decreases according to the same law as the frequency of one of its components, thus from (6)

$$T :: \nu :: \lambda^{-1} :: t^{-1/3} . \qquad (7)$$

The rate of cooling is much slower than in the usual theory, according to which the distance of a galaxy is roughly $:: t$, so that $\lambda :: t$. This makes

$$T :: \nu :: \lambda^{-1} :: t^{-1} . \qquad (8)$$

The observed value of T is about 2.8°K. This gives an energy kT, which may be compared to the rest-energy of a proton to give a dimensionless number

$$kT/m_p c^2 = 2.5 \times 10^{-13} .$$

According to the L.N.H. we should expect this to vary with the epoch according to the law

$$kT/m_p c^2 :: t^{-1/3} .$$

Thus $T :: t^{-1/3}$, in agreement with (7) above, but disagreeing strongly with (8).

The microwave radiation thus provides confirmation of our present picture of the recession of the galaxies. The radiation has been cooling according to the $t^{-1/3}$ law since a time close to the Big Bang. According to the usual views it has been cooling according to the t^{-1} law since a certain decoupling time, when it became decoupled from matter. This decoupling time must have been around $t = 10^{26}$, when T was 10^{13} times greater than now, so that kT was approximately $m_p c^2$. The existence of such a decoupling time, playing a fundamental role in cosmology, would contradict the L.N.H.

We must now alter our view about the age of the Universe. If the distance of a galaxy varies in proportion to f(t), the Hubble constant is

$$H = f'(t)/f(t)$$

With a constant rate of expansion, $f(t) :: t$ and $H = t^{-1}$. With the new theory $f(t) :: t^{1/3}$ and $H = 1/3t^{-1}$. So with the same value for the Hubble constant the age of the Universe is reduced by a factor 3. The presently accepted value for the Hubble constant corresponds to an age of the Universe of about 18×10^9 years with a constant rate of expansion. It thus corresponds to about 6×10^9 years with the new law of retardation of the galaxies. This is rather less than the age one usually believes, but it is not impossible.

PLANETARY ORBITS

Let us see how the variation of G will affect the motion of planets around the sun. We take circular orbits for simplicity. From Newton's law

$$G_A M_A = v_A^2 r_A,$$

where M denotes the mass of the sun and v and r the velocity of the planet and radius of the orbit. The suffix A denotes that the quantities refer to atomic units.

We have $G_A :: t^{-1}$. With conservation of mass, M_A is constant. v_A is dimensionless, a fraction of the velocity of light, so we may drop the suffix A from it. We get

$$v^2 r_A :: t^{-1} . \qquad (9)$$

We have just this one equation, which is inadequate to determine how v and r_A vary with t. Some further assumption is needed.

The time taken by the planet to complete one orbit is

$$\frac{2\pi r_A}{v} .$$

This time is, by definition, one unit of ephemeris time, the time used by astronomers. It is the time to be used in equations of motion, either Newton's or Einstein's. Thus if τ denotes ephemeris time

$$\frac{dt}{d\tau} :: \frac{r_A}{v} . \qquad (10)$$

The possibility that ephemeris time differs from atomic time was first considered by E. A. Milne. He proposed the connection between them

$$\tau = \log t. \qquad (11)$$

This is quite a nice relation. It has the effect that, as one goes back to the Big Bang, $\tau \to 0$, we have $\tau \to -\infty$, so there is plenty of ephemeris time available for the development of star clusters and galaxies.

If one adopts (11), one gets from (10)

$$\frac{r_A}{v} :: t$$

and so from (9)

$$r_A :: t^{1/3} \qquad v :: t^{-2/3} . \tag{12}$$

Thus the planets are all spiralling outward. This is a cosmological effect, superposed on all other effects from known physical causes.

The arguments that Milne used[2,3,4] for setting up his fundamental relation (11) were of a general philosophical character and I find them unconvincing. Can one set up an alternative theory based on more usual physical ideas?

There is such a theory, based on Einstein's theory of gravitation. It involves assuming that the planets move along geodesics in a certain Riemann space. We must not assume that this Riemann space satisfies the Einstein field equations $R_{\mu\nu} = 0$ accurately, as we would then be forced to a theory with constant G. But we may assume that the field equations hold to an accuracy of t^{-1} and contain errors only of order t^{-2}. Such errors are so excessively small that we need not be disturbed by them.

The method of calculation involves assuming a Schwarzschild metric in which there is a multiplying factor that depends on t, and in which also the mass parameter is allowed to vary with t. We are led to the result that, for a circular orbit, v = constant. Thus (9) gives

$$r_A :: t^{-1} , \tag{13}$$

which is very different from the consequences (12) of Milne's assumption. It means that the planets are spiralling inward.

There is now a different connection between ephemeris time and atomic time. From (10) we get

$$\frac{dt}{d\tau} :: t^{-1}$$

leading to

$$\tau :: t^2 . \tag{14}$$

This means that atomic clocks are slowing down with reference to ephemeris time, whereas with Milne's relation (11) they are speeding up.

COMPARISON WITH OBSERVATION

It should be possible to check on these theories with sufficiently accurate astronomical observations. To make the comparison of ephemeris time with atomic time one may use the motion of the moon. The moon has been observed accurately in ephemeris time for some centuries and has been observed with atomic clocks since 1955. Van Flandern has been working for some years on this comparison. The calculations are complicated because of the large tidal distrubances of the moon's motion. The lunar angular acceleration $\dot{n}$ in ephemeris time is produced entirely by tidal and similar mechanical effects. In atomic time it contains an additional cosmological term coming from the difference of atomic and ephemeris time. Thus the difference

$$\dot{n}_A - \dot{n}_{eph}$$

gives the cosmological term.

In a conference that we had here two years ago Van Flandern gave a figure for this difference

$$-10 \text{ seconds}/(\text{century})^2.$$

The minus sign here indicates that the moon is spiralling outward, which agrees with Milne's relation. The magnitude of the effect is about a third of what it should be according to the present theory, when one takes into account the reduction in the age of the Universe.

I recently had a letter from Van Flandern where he says that during the intervening two years the accuracy of his result has not increased in the expected way and he is wondering whether there is not some undiscovered systematic error. So the problem is not at all settled and there is still a possibility that the Einstein relations (13) (14) may turn out to be the correct ones.

Another chance for checking the theory is provided by direct observations of the distance of Mars with the help of transponders that were put on the surface of Mars by the Viking expeditions. After a complete Martian orbit, about two earth years, one should be able to see fairly accurately whether Mars is spiralling outward or inward.

REFERENCES

1. P.A.M Dirac, Proc. Roy Soc. A, 165, 199 (1938).
2. E. A. Milne, Proc. Roy. Soc. A, 158, 324 (1936).
3. E. A. Milne, Proc. Roy. Soc. A, 159, 171 (1937).
4. E. A. Milne, Proc. Roy. Soc. A, 159, 526 (1937).
5. J. Van Flandern, Measurements of Cosmological Variations of the Gravitational Constant, pages 21-28 (1978).

CONCLUDING REMARKS

Eugene P. Wigner
Princeton University, Princeton, NJ 08540

I know I should not say it, but I'll say it nevertheless, that I appreciate very much the invitation to this conference, not only because I want to participate in the honoring of Paul Dirac, but also because of the many interesting addresses, including the last one, which I heard. And as to my own address: it is a pleasure to say a few words about the person I call "my famous brother-in-law." It is not in all regards easy to do this as he does not like to speak about himself - often I believe he does not like to speak. But surely not about himself, his past, his childhood, his parents. But recently he gave an address in Varenna about his scientific development, about his motives in science, and I can recommend to all of us to read that article, to read it twice.

You know, in a way the early careers of Paul and of myself were somewhat similar and even though I surely do not want to compare myself with him, it may be of some interest to compare our scientific developments, the early influences on our thinkings. As many of you may know, we started in a somewhat similar way: both as engineers. He studied electrical engineering, I studied chemical engineering. But the principal effects of these studies were quite different on the two of us. Before the studies, we both were fascinated with the beauty of mathematics, with the ingenuity which was exhibited already by the Greek geometers and which was so tremendously increased later. But what he learned from his engineering studies is that it is not always necessary to have precise results, as a rule not even possible to have precise results, one must often be satisfied with approximations, with the consideration of limiting cases, when some effects can be neglected, and that the results can be beautiful even if one does this. And, as we all know when we review the present status of physics, it consists of approximations, or rather all of it is restricted to the consideration of limiting cases in which some influences are neglected. I could, and would like, to speak at length about this but I should not do it right now. Anyway, what he recalls most vividly from his engineering studies is that he learned that even an incomplete theory can be of immense value, that it can exhibit great beauty - and from a mathematical point of view this should not be surprising - a limiting case, when one of the variables is restricted to zero, or restricted to infinity, may be too simple, but can still

ISSN 0094-243X/78/175/$1.50

have a great deal of beauty and intricacy. After all, Euclidean geometry is a limiting case of Riemannian geometry. Well, as I said, what Paul principally praises as a result of his engineering education is his respect, and surely absence of contempt, for approximations. As to myself, the effect of my engineering studies was quite different. As a chemist, I learned a respect for experimental facts, of experimental facts which are not yet consequences of the theory, but some of the properties of which do follow from the theory, and to respect the theory even if its consequences account for only a small part of the experimental findings. The theory tells us that if we have a precipitate of 214 gr. of $BaSO_4$, the amount of sulfuric acid that was in our solution was 80 gr. - but it does not tell us, or did not tell us when I learned chemistry, why the $BaSO_4$ precipitates. Surely, it does not tell us, and we do not expect it to tell us, why there was some sulphuric acid in the solution.

This difference in our thinking, his respect for approximations and desire to improve them and to add to the beauty of the resulting theory, my respect for experimental facts and for partial explanations of them, stayed with us throughout our careers.

Let me now speak a little more about Paul. We all know about his scientific accomplishments, about his early contributions to the creation of quantum mechanics-Heisenberg also admired this - about the equation, the 50th jubilee we celebrate today, about his creation of the fundaments of quantum field theory. And we heard just now about his new idea, a fascinating idea. I can not resist telling a short story about the discovery of the relativistic electron equation. Jordan and I worked on this question at the same time Paul did but we got stuck. Then one day Born, our director, received a letter from Paul, recommending a number of changes in an article which Born had asked him to review. He recommended the union of chapters 2 and 3, which made the elimination of an equation possible, and he recommended some other changes. At the end he added a paragraph, a single paragraph, telling about his relativistic electron equation. Jordan, who read the letter first, told me, "I wish we had discovered that equation ourselves but it is so beautiful that I am glad it was discovered."

Well, the story I just told demonstrates not only Paul's scientific ingenuity, it also demonstrates his modesty. This is demonstrated also by his love to take Sunday walks, alone - alone and devoid of any recognition by others. Another property which I wish to mention is his inclination to recognise gifts of knowledge and interest that his teachers and friends imparted to him, his inclination to gratitude. When he speaks about his early teachers, about Broad from whom he learned relativity theory, about Fraser who acquainted him with the beauty of projective geometry, about Fowler who awakened his interest in quantum theory, he acknowledges the deserts of them wholeheartedly. His absolute confidence in the simplicity of the world and in our ability to discover the basis of this is another character of Paul - an admirable conviction though not shared by me.

The last property of Paul that I wish to mention is his retiring nature. Once Polanyi, Paul and I had a luncheon together and there was a lively discussion on the purposes and deserts of science. Paul hardly spoke a word so that when we left, I told him, "Paul, why do you not speak up a bit more? People are interested in your views, they want to know what you believe." Paul's reply was characteristic of his incisive thoughtfulness and also of his retiring nature. He said, "There are always more people willing to speak than willing to listen." Well, having told you this, I am now also going to listen and I finish my speech.

Index of Authors

AIP Conference Proceedings

		L.C. Number	ISBN
No.1	Feedback and Dynamic Control of Plasmas (Princeton) 1970	70-141596	0-88318-100-2
No.2	Particles and Fields - 1971 (Rochester)	71-184662	0-88318-101-0
No.3	Thermal Expansion - 1971 (Corning)	72-76970	0-88318-102-9
No.4	Superconductivity in d- and f-Band Metals (Rochester, 1971)	74-18879	0-88318-103-7
No.5	Magnetism and Magnetic Materials - 1971 (2 parts) (Chicago)	59-2468	0-88318-104-5
No.6	Particle Physics (Irvine, 1971)	72-81239	0-88318-105-3
No.7	Exploring the History of Nuclear Physics (Brookline, 1967, 1969)	72-81883	0-88318-106-1
No.8	Experimental Meson Spectroscopy - 1972 (Philadelphia)	72-88226	0-88318-107-X
No.9	Cyclotrons - 1972 (Vancouver)	72-92798	0-88318-108-8
No.10	Magnetism and Magnetic Materials - 1972 (2 parts) (Denver)	72-623469	0-88318-109-6
No.11	Transport Phenomena - 1973 (Brown University Conference)	73-80682	0-88318-110-X
No.12	Experiments on High Energy Particle Collisions - 1973 (Vanderbilt Conference)	73-81705	0-88318-111-8
No.13	π-π Scattering - 1973 (Tallahassee Conference)	73-81704	0-88318-112-6
No.14	Particles and Fields - 1973 (APS/DPF Berkeley)	73-91923	0-88318-113-4
No.15	High Energy Collisions - 1973 (Stony Brook)	73-92324	0-88318-114-2
No.16	Causality and Physical Theories (Wayne State University, 1973)	73-93420	0-88318-115-0
No.17	Thermal Expansion - 1973 (Lake of the Ozarks)	73-94415	0-88318-116-9
No.18	Magnetism and Magnetic Materials - 1973 (2 parts) (Boston)	59-2468	0-88318-117-7
No.19	Physics and the Energy Problem - 1974 (APS Chicago)	73-94416	0-88318-118-5
No.20	Tetrahedrally Bonded Amorphous Semiconductors (Yorktown Heights, 1974)	74-80145	0-88318-119-3
No.21	Experimental Meson Spectroscopy - 1974 (Boston)	74-82628	0-88318-120-7
No.22	Neutrinos - 1974 (Philadelphia)	74-82413	0-88318-121-5
No.23	Particles and Fields - 1974 (APS/DPF Williamsburg)	74-27575	0-88318-122-3
No.24	Magnetism and Magnetic Materials - 1974 (20th Annual Conference, San Francisco)	75-2647	0-88318-123-1
No.25	Efficient Use of Energy (The APS Studies on the Technical Aspects of the More Efficient Use of Energy)	75-18227	0-88318-124-X

No.26	High-Energy Physics and Nuclear Structure - 1975 (Santa Fe and Los Alamos)	75-26411	0-88318-125-8
No.27	Topics in Statistical Mechanics and Biophysics: A Memorial to Julius L. Jackson (Wayne State University, 1975)	75-36309	0-88318-126-6
No.28	Physics and Our World: A Symposium in Honor of Victor F. Weisskopf (M.I.T., 1974)	76-7207	0-88318-127-4
No.29	Magnetism and Magnetic Materials - 1975 (21st Annual Conference, Philadelphia)	76-10931	0-88318-128-2
No.30	Particle Searches and Discoveries - 1976 (Vanderbilt Conference)	76-19949	0-88318-129-0
No.31	Structure and Excitations of Amorphous Solids (Williamsburg, Va., 1976)	76-22279	0-88318-130-4
No.32	Materials Technology - 1975 (APS New York Meeting)	76-27967	0-88318-131-2
No.33	Meson-Nuclear Physics - 1976 (Carnegie-Mellon Conference)	76-26811	0-88318-132-0
No.34	Magnetism and Magnetic Materials - 1976 (Joint MMM-Intermag Conference, Pittsburgh)	76-47106	0-88318-133-9
No.35	High Energy Physics with Polarized Beams and Targets (Argonne, 1976)	76-50181	0-88318-134-7
No.36	Momentum Wave Functions - 1976 (Indiana University)	77-82145	0-88318-135-5
No.37	Weak Interaction Physics - 1977 (Indiana University)	77-83344	0-88318-136-3
No.38	Workshop on New Directions in Mössbauer Spectroscopy (Argonne, 1977)	77-90635	0-88318-137-1
No.39	Physics Careers, Employment and Education (Penn State, 1977)	77-94053	0-88318-138-X
No.40	Electrical Transport and Optical Properties of Inhomogeneous Media (Ohio State University, 1977)	78-54319	0-88318-139-8
No.41	Nucleon-Nucleon Interactions - 1977 (Vancouver)	78-54249	0-88318-140-1
No.42	Higher Energy Polarized Proton Beams (Ann Arbor, 1977)	78-55682	0-88318-141-X
No.43	Particles and Fields - 1977 (APS/DPF, Argonne)	78-55683	0-88318-142-8
No.44	Future Trends in Superconductive Electronics (Charlottesville, 1978)	77-9240	0-88318-143-6
No.45	New Results in High Energy Physics - 1978 (Vanderbilt Conference)	78-67196	0-88318-144-4
No.46	Topics in Nonlinear Dynamics (La Jolla Institute)	78-057870	0-88318-145-2
No.47	Clustering Aspects of Nuclear Structure and Nuclear Reactions (Winnepeg, 1978)	78-64942	0-88318-146-0
No.48	Current Trends in the Theory of Fields (Tallahassee, 1978)	78-72948	0-88318-147-9